introduction to

REAL
ANALYSIS

MICHAEL GEMIGNANI
Associate Professor of Mathematics, Smith College

1971

W. B. SAUNDERS COMPANY · PHILADELPHIA · LONDON · TORONTO

W. B. Saunders Company: West Washington Square
Philadelphia, Pa. 19105

12 Dyott Street
London WC1A   1DB

1835 Yonge Street
Toronto 7, Ontario

Introduction to Real Analysis   SBN 0-7216-4095-8

Print No.:   9   8   7   6   5   4   3   2   1

To Mom and Dad Federico

PREFACE

My purpose in writing this short book has been to present in a well-motivated and natural sequence the basic ideas of classical real analysis. My intention has been to offer a book that can be covered from beginning to end in one semester. With this objective in mind, and exercising as much control as possible over my own prejudices, I have sought to pare away all material that might be considered extraneous in an introductory treatment of real analysis. Nevertheless, enough material is covered to provide a firm base on which to build for later studies in multivariate calculus and complex analysis. The only background material required of the reader is two semesters of standard calculus. Otherwise the text is self-contained.

The book starts with some preliminaries on sets and functions, followed by a quick review of the essential properties of the integers. The substance of the book begins in Chapter 2, where, using Dedekind cuts, the set of real numbers is constructed. This foundation supports the subsequent chapters: topological framework; sequences and series of real numbers; differentiation; Riemann integration, with a quick look at the Riemann-Stieltjes integral; and, finally, sequences and series of functions.

The tone of the book is classical. No attempt is made to present important concepts and results of analysis in any setting more general than that of the real numbers. However, some terminology has been updated and modern methods applied to smooth out and shorten certain classical techniques.

In the editorial development of the book I received valuable help from Professors George E. Andrews of Pennsylvania State University, John E. Brothers of Indiana University, Robert B. Burckel of the University of Oregon, Rene F. Dennemeyer of California State College at San Bernardino, Richard E. Dowds of SUNY College at Fredonia, and Frederick Greenleaf of New York University, each of whom critically reviewed the entire manuscript. Their detailed and percipient comments resulted in many improvements to the final manuscript.

MICHAEL GEMIGNANI

CONTENTS

4

5

6

7

1

PRELIMINARIES

1.1 SETS AND FUNCTIONS

SETS

It is assumed that any reader of this book has already had some experience with sets; hence most of what is said in this section will be for the sake of review rather than for the purpose of presenting new material.

A *set* will be taken to be any clearly defined collection of objects; the objects in a set are called *elements*, or *points*, of the set. If x is an element of the set S, we write $x \in S$. We denote the phrase *is not an element of* by $\notin$.

Sets may be denoted either by explicitly listing their elements inside of braces (for example, $\{a, b, c\}$ is the set having a, b, and c as elements) or by giving the property that every element of the set possesses and which is possessed by no element not in the set, in this format: $\{x \mid$ property x must satisfy to be in set$\}$. For example, $\{x \mid x$ is a barn$\}$ is the set of all barns, or the set of all x such that x is a barn.

A set S is said to be a *subset* of a set T if each element of S is an element of T. We usually denote S *is a subset of* T by $S \subseteq T$. Two sets are *equal* if they contain exactly the same elements; that is, $S = T$ if $S \subseteq T$ and $T \subseteq S$. The phrase *is not a subset of* is denoted by $\nsubseteq$. S is said to be a *proper subset* of T if $S \subseteq T$ but $S \neq T$. We denote S *is a proper subset of* T by $S \subset T$.

The empty set, that is, the set which contains no elements whatsoever, is denoted by $\varnothing$.

If S and T are any two sets, then the *complement of S in T* is the set of all elements of T which are not elements of S; we denote the complement of S in T by $T \sim S$.

The two most basic set operations are *union* and *intersection*. If $\{S_i\}$, $i \in I$, is any family of sets indexed by some set I (the reader may consider I simply as a set of labels for distinguishing the various members of the family of sets),

1

then the *union* of this family of sets is defined to be $\{x \mid x \in S_i$ for at least one $i \in I\}$. The union of $\{S_i\}$, $i \in I$, may be denoted by $\bigcup_I S_i$, or $\bigcup \{S_i \mid i \in I\}$. The *intersection* of this family of sets, denoted by $\bigcap_I S_i$, or $\bigcap \{S_i \mid i \in I\}$, is defined to be $\{x \mid x \in S_i$ for every $i \in I\}$. When only a few sets are involved, say S_1, S_2, and S_3, the union and intersection of these sets may be denoted by $S_1 \cup S_2 \cup S_3$ and $S_1 \cap S_2 \cap S_3$, respectively.

We assume that the reader is already moderately familiar with these set operations, at least so far as any finite family of sets is concerned. We now prove the *DeMorgan formulas* for an arbitrary family of sets.

Proposition 1:   Let $\{S_i\}$, $i \in I$, be any family of subsets of some set T. Then

a) $\bigcup_I (T \sim S_i) = T \sim \bigcap_I S_i$, and

b) $\bigcap_I (T \sim S_i) = T \sim \bigcup_I S_i$.

PROOF: We prove (a) and leave the proof of (b) as an exercise. To prove that two sets are equal we must show that they contain the same elements. Suppose $x \in \bigcup_I (T \sim S_i)$; then $x \in T \sim S_i$ for at least one $i \in I$. Therefore $x \notin S_i$ for at least one $i \in I$; hence $x \notin \bigcap_I S_i$. Consequently, $x \in T \sim \bigcap_I S_i$. We have thus proved that every element of $\bigcup_I (T \sim S_i)$ is an element of $T \sim \bigcap_I S_i$; that is,

$$\bigcup_I (T \sim S_i) \subseteq T \sim \bigcup_I S_i. \tag{1}$$

Suppose now that $x \in T \sim \bigcap_I S_i$. Then there is some $i \in I$ for which $x \notin S_i$ (or else x would be an element of $\bigcap_I S_i$). Therefore $x \in T \sim S_i$ for some $i \in I$. Then $x \in \bigcup_I (T \sim S_i)$. Consequently,

$$T \sim \bigcap_I S_i \subseteq \bigcup_I (T \sim S_i). \tag{2}$$

Combining (1) and (2) we have the equality in (a).

If S and T are any two sets, then the *Cartesian product* of S and T is defined to be the set of all *ordered pairs* (s, t) such that $s \in S$ and $t \in T$. The Cartesian product of S and T is denoted by $S \times T$. We say that two ordered pairs (s, t) and (s', t') are equal and write $(s, t) = (s', t')$ if $s = s'$ and $t = t'$. (More formally, one can define the ordered pair (s, t) to be the set $\{\{s\}, \{s, t\}\}$; the properties of ordered pairs can then be developed set-theoretically.)

If S and T are any sets, then a subset R of $S \times T$ is said to be a *relation between S and T*. A subset of $S \times S$ is said to be a *relation on S*. If R is a relation between S and T, that is, $R \subseteq S \times T$, then if $(s, t) \in R$, we may also write sRt or say that s and t are *R-related*. Although, strictly speaking, a relation is a set, at times a phrase or symbol defining the relation will be used in place of the actual set. For example, although *is equal to* defines a relation on the collection of subsets of some set, we usually write simply $S = T$ if S and T

are equal subsets, rather than explicitly refer to any relation. It should be noted that a relation R between S and T is not necessarily a relation between T and S; nor is it necessary that each element of S be R-related to some element of T or that each element of T be R-related to some element of S.

FUNCTIONS

A *function f* from a set S into a set T is a relation between S and T such that each element of S is f-related to one and only one element of T. If $(s, t) \in f$, we may write $t = f(s)$. Less formally, a function f from S into T is a rule, or other device, by which each element s of S is corresponded with precisely one element $f(s)$ of T. Functions are usually defined by giving a rule which enables us to find $f(s)$ whenever s is given. Again, rarely is explicit mention made of the fact that a function is a set. Functions are often called *maps* or *mappings*.

We may write $f:S \to T$ to denote that f is a function from S into T. If $f:S \to T$, we call S the *domain* and T the *range* of f; $\{t \in T \mid t = f(s)$ for some $s \in S\}$ is said to be the *image* of f.* The image of f may be denoted by $f(S)$.

Example 1: Suppose $f:S \to T$ and A and B are both subsets of S. We now prove that $f(A \cup B) \subseteq f(A) \cup f(B)$. Let $y \in f(A \cup B)$. Then $y = f(x)$ for some $x \in A \cup B$. Now $x \in A$ or $x \in B$; consequently, $f(x) \in f(A)$ of $f(x) \in f(B)$. In either case, $f(x) = y \in f(A) \cup f(B)$. Therefore it follows that $f(A \cup B) \subseteq f(A) \cup f(B)$.

Suppose that $y \in f(A) \cup f(B)$. Then $y \in f(A)$ or $y \in f(B)$. This implies there is $x \in A \cup B$ such that $f(x) = y$; hence, $y \in f(A \cup B)$. Thus, $f(A) \cup f(B) \subseteq f(A \cup B)$; hence, $f(A) \cup f(B) = f(A \cup B)$.

If $f:S \to T$ and $f(S) = T$, we say that f is *onto;* that is, f is onto T if each element of T is a *function value* at least once. If $f(s) = f(s')$ implies $s = s'$ for any $s, s' \in S$, then f is said to be *one-one;* that is, f is one-one if each element of T appears as a function value at most once. The term *injective* is sometimes used for *one-one*, and *surjective* is sometimes used for *onto*. An *injection* is a one-one function, while a *surjection* is an onto function. A function which is both one-one and onto is sometimes called a *bijection*.

Suppose $f:S \to T$. For each $t \in T$, we define $f^{-1}(t)$ to be $\{s \in S \mid f(s) = t\}$. For some $t \in T$, $f^{-1}(t)$ may be empty; likewise, $f^{-1}(t)$ may contain finitely or infinitely many elements. If for each $t \in T$ $f^{-1}(t)$ contains precisely one element of S (that is, if for each $t \in T$ there is precisely one $s \in S$ such that $f(s) = t$), then we can consider f^{-1} as defining a function from T onto S; that is, $f^{-1}(t) = s$, where s is that unique element of S such that

---

* Some authors define both the range and the image of f to be $\{t \in T \mid t = f(s)$ for some $s \in S\}$.

$f(s) = t$, for each $t \in T$. The condition under which f^{-1} defines a function from T into S is given in the following proposition.

Proposition 2:   If $f: S \to T$, then f^{-1} defines a function from T into S, if and only if f is one-one and onto.

PROOF: Suppose f is both one-one and onto. Then given any $t \in T$, there is one (onto) and only one (one-one) element $s \in S$ such that $f(s) = t$. Therefore $f^{-1}(t) = s$ is uniquely defined for each $t \in T$; hence f^{-1} is a function from T into S.

Suppose that f^{-1} is a function from T into S. Then for each $t \in T$, $f^{-1}(t) = s$, s a unique element of S. Therefore no element of T is the image of more than one element of S; hence f is one-one. Moreover, since $f^{-1}(t) = s$ implies $f(s) = t$ and $f^{-1}(t)$ is an element of S for each $t \in T$, f is also onto. Therefore f is one-one and onto.

If $f: S \to T$ and $U \subseteq T$, we define $f^{-1}(U)$ to be $\{s \in S \mid f(s) \in U\}$.

Example 2:   Consider the function $f: \mathbb{R} \to \mathbb{R}$, $\mathbb{R}$ the set of real numbers, defined by $f(x) = 2x$ for each $x \in \mathbb{R}$. Then $f^{-1}(y) = y/2$ for each $y \in \mathbb{R}$ since $f(y/2) = y$. Clearly, f^{-1} is also a function from $\mathbb{R}$ into $\mathbb{R}$, as we could have deduced from the fact that f is both one-one and onto. If $U = \{x \mid 0 < x \leqslant 1\}$, then $f^{-1}(U) - \{y \mid 0 < y \leqslant 1/2\}$.

Suppose $f: S \to T$ and $g: T \to W$. Then $g \circ f$, the *composition of* g *with* f, is a function from S into W defined by $(g \circ f)(s) = g(f(s))$ for each $s \in S$.

EXERCISES

1. Express in words each of the following:

a) $1 \in \{1, 2, 3\}$ 　　　　 d) $\{x \mid x^2 = 1\}$
b) $\{q \mid q \text{ is an integer}\}$ 　 e) $\varnothing \neq \{\varnothing\}$
c) $\varnothing \notin \{\{\varnothing\}\}$ 　　　　 f) $S \cap T \subseteq S \cup T$

2. Prove each of the following:

a) If $S \subseteq T$, then $T \sim (T \sim S) = S$.
b) $\varnothing \subseteq S$ for any set S.
c) If T is any set and $\{U_i\}$, $i \in I$, is a family of subsets of T, then $T \cap (\bigcup_I U_i) = \bigcup_I (T \cap U_i)$.
d) If $f: S \to T$ and $\{U_i\}$, $i \in I$, is a family of subsets of T, then $f^{-1}(\bigcup_I U_i) = \bigcup_I f^{-1}(U_i)$.
e) If $f: S \to T$ and $A \subseteq T$, then $f(f^{-1}(A)) \subseteq A$.
f) If $f: S \to T$ and $\{U_i\}$, $i \in I$, is a family of subsets of T, then $f^{-1}(\bigcap_I U_i) = \bigcap_I f^{-1}(U_i)$.

3. Prove (b) of Proposition 1.

4. Suppose $f: S \to T$ and A and B are subsets of S. Prove or disprove:

a) $f(A \cup B) = f(A) \cup f(B)$

b) $f(A \cap B) = f(A) \cap f(B)$

1.2 PARTIAL AND TOTAL ORDERINGS. EQUIVALENCE RELATIONS

PARTIAL ORDERINGS

If $\mathbb{R}$ is the set of real numbers and $\leqslant$ denotes *less than or equal to*, then $\leqslant$ defines a relation on $\mathbb{R}$ which has the following properties:

P1) $x \leqslant x$ for any $x \in \mathbb{R}$;
P2) $x \leqslant y$ and $y \leqslant x$ implies $x = y$ for any $x, y \in \mathbb{R}$;
P3) $x \leqslant y$ and $y \leqslant z$ implies $x \leqslant z$ for any x, y, and z in $\mathbb{R}$;
T) $x, y \in \mathbb{R}$ implies $x \leqslant y$ or $y \leqslant x$.*

Any relation on a set S which shares properties (P1), (P2), and (P3) is said to be a *partial ordering* of S. Any partial ordering of S which also satisfies (T) is said to be a *total ordering* of S. We may denote a set with partial ordering $\leqslant$ by $S, \leqslant$.

Example 3: Let $P(S)$ be the family of all subsets of some set S; we call $P(S)$ the *power set* of S. Then $P(S)$ is partially ordered by $\subseteq$. In particular (P1) to (P3) become:

P1) If W is any subset of S, then $W \subseteq W$.

P2) If W and T are subsets of S such that $W \subseteq T$ and $T \subseteq W$, then $W = T$.

P3) If W, T, and Z are subsets of S such that $W \subseteq T$ and $T \subseteq Z$, then $W \subseteq Z$.

Since for any two subsets of S, one is not necessarily a subset of the other, $\subseteq$ is not a total ordering of $P(S)$.

Let $S, \leqslant$ be any partially ordered set, and let W be any subset of S. An element u of S is said to be an *upper bound* for W if $w \leqslant u$ for each $w \in W$. An element v of S is said to be a *lower bound* for W if $v \leqslant w$ for each $w \in W$. It is not necessarily true that every non-empty subset of a partially ordered set has either an upper bound or a lower bound.

---

* These properties are presented here informally in that they are not proved, but it is assumed that anyone who has worked with the real numbers will immediately recognize them as being true. When we define the real numbers in Chapter 2 we will define $\leqslant$ formally and prove these given properties.

Example 4: Let $\mathbb{Z}$ be the set of integers. Then $\mathbb{Z}$ is totally ordered by $\leqslant$. $\mathbb{Z}$ has neither an upper bound nor a lower bound. The set $\mathbb{N}$ of positive integers has 0, -1, -2, and so forth, as lower bounds in $\mathbb{Z}$, but $\mathbb{N}$ has no upper bound. The set of negative integers has upper bounds, but no lower bounds. Any finite set of integers has both upper and lower bounds in $\mathbb{Z}$. In this case, any set which has an upper bound (lower bound) will have an infinite number of such bounds.

Let W be a subset of a partially ordered set S, $\leqslant$. Then an element u' of S is said to be a *least upper bound* for W if u' is an upper bound for W and $u' \leqslant u$ for any upper bound u of W. Any element v' of S is said to be the *greatest lower bound* of W if v' is a lower bound for W and $v' \leqslant v$ for any lower bound v of W. The least upper bound and greatest lower bound for W may be denoted by lub W and glb W, respectively. One can easily show that $u' = $ lub W if and only if $u' \in S$ is an upper bound for W and given any $x \in W$, $x < u'$, there is $y \in W$ such that $x < y \leqslant u'$.

It is not necessarily true in an arbitrary partially ordered set that any set which has an upper bound (lower bound) has a least upper bound (greatest lower bound). Nor is it true that any set which has a least upper bound (greatest lower bound) necessarily contains that bound as an element.

Example 5: Let $\mathbb{R}$ be the set of real numbers totally ordered by $\leqslant$. Then $\{x \mid x < 1\}$ has 1 as its least upper bound, but does not contain 1.

THE POSITIVE INTEGERS

The positive integers can be characterized (defined) as a totally ordered set $\mathbb{N}$ having the following properties:

N1) $\mathbb{N}$ contains a least element; this least element is denoted by 1.

N2) For each $n \in \mathbb{N}$, there is an *immediate successor* of n: the immediate successor of n is denoted by $n + 1$. (Formally, $n + 1 = $ glb $\{x \in \mathbb{N} \mid n < x\}$.)

N3) (*Principle of Finite Induction*) If a subset T of $\mathbb{N}$ contains 1 and T contains $n + 1$ whenever $n \in T$, then $T = \mathbb{N}$.

All the order and algebraic properties of $\mathbb{N}$ can be developed from (N1), (N2), and (N3).

The following is an example of a proof by finite induction (N3).

Example 6: We will prove that the sum of the first n positive integers is equal to $(n/2)(n + 1)$. Let T be the set of positive integers n for which the sum of the first n positive integers is given by

$$1 + 2 + 3 + \ldots + n = (n/2)(n + 1). \tag{3}$$

Direct computation, $1 = (1/2)(2)$, shows that $1 \in T$. Suppose $n \in T$. Then the sum of the first $n + 1$ positive integers is $1 + 2 + \ldots + n + (n + 1) = (1 + 2 + 3 + \ldots + n) + (n + 1) = (n/2)(n + 1) + (n + 1)$ (since $n \in T$, and hence (3) is applicable to n) $= (n + 1)(n/2 + 1) = ((n + 2)/2)(n + 1) = ((n + 1)/2)(n + 2)$, which is the appropriate expression for $n + 1$. Therefore if $n \in T$, $n + 1 \in T$; hence by (N3), $T = \mathbb{N}$. Consequently, (3) is true for any positive integer n.

From an order point of view, the set of integers can be characterized as a totally ordered set $\mathbb{Z}$, $\leqslant$ such that for any $z \in \mathbb{Z}$ $\{x \in \mathbb{Z} \mid z < x\}$ satisfies (N1) to (N3) and $\mathbb{Z}$ has no lower bound.

EQUIVALENCE RELATIONS

A partial ordering is an example of a special kind of relation that can be defined on a set. Another important type of relation is an *equivalence relation*. The prototype for an equivalence relation is $=$. Since ambiguity is likely to result if $=$ is used to denote an arbitrary equivalence relation, E will be used instead. A relation E on a set S is said to be an *equivalence relation* on S if E satisfies the following properties:

E1) $s \, E \, s$ for any $s \in S$.
E2) If s and s' are any elements of S such that $s \, E \, s'$, then $s' \, E \, s$.
E3) If s, s', and s'' are any elements of S such that $s \, E \, s'$ and $s' \, E \, s''$, then $s \, E \, s''$.

Compare (E1) through (E3) with the properties of $=$. Note that the only difference between a partial ordering on S and an equivalence relation on S is that property (P2) has been replaced by property (E2).

Example 7: Let T be the set of all plane triangles. Then *is similar to* defines an equivalence relation on T. An equivalence relation on T is also defined by *is congruent to;* still another equivalence relation on T is defined by *has the same area as*.

The most important property of an equivalence relation is given in the following proposition.

Proposition 3: Let S be any set, and let E be an equivalence relation on S. For each $s \in S$, define $\bar{s} = \{x \in S \mid x \, E \, s\}$. We call $\bar{s}$ the *equivalence class* of s. Then each element in S is contained in one and only one equivalence class.

PROOF: Since by (E1) we have $s \, E \, s$ for each $s \in S$, $s \in \bar{s}$; therefore each element of S is in at least one equivalence class. We now prove that if two equivalence classes are not disjoint (that is, have an empty intersection), then they must be equal. Suppose that y is an element of both $\bar{s}$ and $\bar{t}$ for some elements s and t of S. Since $y \in \bar{s} \cap \bar{t}$, we have both $y \, E \, s$ and $y \, E \, t$. If $x \in \bar{s}$, then we have $x \, E \, s$. Since $y \, E \, s$ implies $s \, E \, y$, we have $x \, E \, s$ and $s \, E \, y$ which implies $x \, E \, y$. But $x \, E \, y$ and $y \, E \, t$ implies $x \, E \, t$, or $x \in \bar{t}$. Therefore $\bar{s} \subseteq \bar{t}$. A similar argument shows $\bar{t} \subseteq \bar{s}$. Therefore $\bar{s} = \bar{t}$.

EXERCISES

1. Let $\mathbb{N}$ be the set of positive integers. Define the relation E on $\mathbb{N} \times \mathbb{N}$ by $(n, m)E(n', m')$ if $n + m' = n' + m$. Prove that E is an equivalence relation on $\mathbb{N} \times \mathbb{N}$. Let $\mathbb{Z}$ be the set of equivalence classes and denote the equivalence class of (n, m) by $|n, m|$. Define $|n, m| + |n' \, m'| = |n + n', m + m'|$ and $|n, m| \cdot |n', m'| = |nn' + mm', mn' + m'n|$, for any $|n, m|$ and $|n', m'|$ in $\mathbb{Z}$. Prove each of the following:

a) If $|n'', m''| = |n, m|$, then $|n'', m''| + |n', m'| = |n, m| + |n', m'|$ and $|n'', m''| \cdot |n', m'| = |n, m| \cdot |n', m'|$; that is, addition and multiplication of the equivalence classes is well-defined (depends only on the equivalence class and not on the representative of the equivalence class).

b) $|n, m| + |a, a| = |n, m|$ for any $|n, m| \in \mathbb{Z}$, where a is any element of $\mathbb{N}$.

c) $|n, m| + |m, n| = |a, a|$.

If one thinks of $|n, m|$ as representing the *integer* $n - m$ and $\mathbb{Z}$ as the set of integers, what further properties would be expected of $\mathbb{Z}$? Prove some of these properties. We have thus constructed the set of integers $\mathbb{Z}$ from $\mathbb{N}$.

2. Let $\mathbb{N}$ be the set of positive integers (considered informally). Let $n \mid m$ denote n divides m, that is, $m = nk$ for some positive integer k. Prove that $\mid$ defines a partial ordering on $\mathbb{N}$. Prove that every two-element subset of $\mathbb{N}$ has a least upper bound and greatest lower bound with respect to $\mid$. (Do not confuse the partial orderings $\mid$ and $\leq$.)

3. Prove that every non-empty subset of the positive integers contains a least element.

4. Find all equivalence relations on a set S which are also partial orderings of S.

5. Let $\mathbb{N}$ be the set of positive integers. We define a relation E on $\mathbb{N} \times \mathbb{N}$ as follows: Let $(n, m) \, E \, (n', m')$ if

$nm' = mn'$. Prove that E is an equivalence relation on $\mathbb{N} \times \mathbb{N}$.

6. Prove, using finite induction, that the sum of the first n squares, that is, $1^2 + 2^2 + \ldots + n^2$, is given by

$$(1/6)(n) \times (n + 1)(2n + 1).$$

7. Prove that if $W \subseteq S$, then $u' = \text{lub } W$ if and only if $u' \in S$ is an upper bound for W and given any $x \in W$, $x < u'$, there is $y \in W$ such that $x < y \leqslant u'$.

8. Given integers n and m we say that m is congruent to n modulo p, p some non-zero integer, if $n - m$ is divisible by p. Prove that *is congruent to modulo p* defines an equivalence relation on the set of integers. How many equivalence classes are there relative to p?

1.3 CARDINALITY

We say that two sets S and T have the *same number of elements*, or the *same cardinality*, if there is a one-one function f from S onto T. That is, S and T have the same cardinality if the elements of S can be put in one-one correspondence with all the elements of T.

A set S is said to be *finite* if S has the same cardinality as $\varnothing$ or if there is a positive integer n such that S has the same number of elements as $\{1, 2, \ldots, n\}$. If S is not finite, we say that S is *infinite*. A set S is said to be *countable* if S has the same cardinality as a subset of $\mathbb{N}$, the set of positive integers.* Otherwise, S is said to be *uncountable*. Thus any finite set is certainly countable.

Proposition 4

a) Any subset of a finite set is finite.
b) Any subset of any countable set is countable.

PROOF: (a) Since S is finite, either $S = \varnothing$ or there is a positive integer n such that S has the same cardinality as $\{1, 2, \ldots, n\}$. If $S = \varnothing$, then the only subset of S is $\varnothing$, which is finite. Assume $S \neq \varnothing$. Then there is a one-one function from S onto $\{1, 2, \ldots, n\}$ for an appropriate n. Suppose $W \subseteq S$. If $W = \varnothing$, then W is finite. If $W \neq \varnothing$, let $i_1, i_2, \ldots, i_m$ be

---

* The term *denumerable* is sometimes used as a synonym for *countable;* some texts also use denumerable to designate countable and infinite.

the elements of $\{1, 2, \ldots, n\}$ in the image of W. Then defining $g:W \rightarrow \{1, 2, \ldots, m\}$ by $g(w) = j$, where $f(w) = i_j$, for each $w \in W$, we see that W is finite.

We leave the proof of (b) as an exercise.

Proposition 5: Let $\{A_n\}$, $n \in \mathbb{N}$, be a countable collection of countable sets. Then $\bigcup_\mathbb{N} A_n$ is also countable ($\mathbb{N}$ represents the set of positive integers).

PROOF: We may enumerate the elements of A_j as $a_{j1}, a_{j2}, a_{j3}, \ldots$, for each positive integer j, but not counting any element twice; that is, when some $a \in A_j$ is reached which is an element of A_i, $i < j$, then a is simply not enumerated again. Thus a_{ji} is the "ith" element of A_j. Let f be defined by $f(a_{ij}) = 3^i 5^j$. Then f is a one-one function from $\bigcup_\mathbb{N} A_n$ onto a subset of $\mathbb{N}$. Therefore $\bigcup_\mathbb{N} A_n$ must be countable.

Proposition 6: If A and B are countable sets, then $A \times B$ is countable.

PROOF: Let $a_1, a_2, \ldots$ and $b_1, b_2, \ldots$ be enumerations of the elements of A and B, respectively. Then each element of $A \times B$ has the form (a_i, b_j) for positive integer i and j. Let $f:A \times B \rightarrow \mathbb{N}$ be defined by $f(a_i, b_j) = 3^i 5^j$. Since f is one-one, $A \times B$ is countable.

Proposition 7: The set $\mathbb{Z}$ of integers is countable.

PROOF: The following scheme indicates a one-one function f from $\mathbb{Z}$ onto $\mathbb{N}$:

$f(z)$	1	2	3	4	5	6	7
z	0	1	-1	2	-2	3	-3

More formally, f is defined by $f(z) = 2z$ if z is positive, and $f(z) = (-1)(2z) + 1$ if z is non-positive.

Corollary: The set $\mathbb{Q}$ of rational numbers is countable. (We view the rational numbers here informally as fractions of the form m/n, m and n integers, $n \neq 0$.)

PROOF: Associating m/n with (m, n), $\mathbb{Q}$ has the same number of elements as a subset of $\mathbb{Z} \times \mathbb{Z}$, which is a countable set by Proposition 6. Therefore $\mathbb{Q}$ is countable.

The following example shows that some sets are uncountable.

Example 8: The set S of unending decimals between 0 and 1 and which contain only 0 and 1 as digits is uncountable. We prove this as follows. Let f be an arbitrary function from $\mathbb{N}$, the set of positive integers, into S. We will show that f cannot be onto. Therefore there will be no one-one function from $\mathbb{N}$ onto S; hence $\mathbb{N}$ and S cannot have the same number of elements.

We form an element $D = .x_1 x_2 x_3 x_4 \ldots$ of S as follows: If the first digit of $f(1)$ is 0, let $x_1 = 1$; but if the first digit of $f(1)$ is 1, let $x_1 = 0$. Therefore D differs from $f(1)$ in the first digit. We choose x_2 to be either 0 or 1, but different from the second digit of $f(2)$. In general, we let x_n be 0 or 1, but different from the nth digit of $f(n)$. Then D is not $f(n)$ for any n since D differs from $f(n)$ for any n at least in the nth digit. Consequently, D is not in the image of f; therefore f is not onto S.

We note that S of Example 8 is a subset of $\mathbb{R}$, the set of real numbers. It follows then that $\mathbb{R}$ is uncountable. For if $\mathbb{R}$ were countable, then S would also be countable since it would then be a subset of a countable set.

EXERCISES

1. Prove (b) of Proposition 4.

2. Let S be a set of more than one element and let S^S denote the set of functions from S into S. Prove that S and S^S do not have the same number of elements. Hint: If S and S^S have the same number of elements, then there is a function f from S onto S^S; in this case, $f(s):S \to S$. Define a function $g:S \to S$ so that $g(s) \neq (f(s))(s)$ for any $s \in S$. (Why is this possible?) Show that $g \neq f(s)$ for any $s \in S$; hence f cannot be onto.

3. Let $P(S)$ be the set of all subsets of the set S. Prove that S and $P(S)$ do not have the same number of elements. (Hint: If S has the same number of elements as $P(S)$, there is a one-one function f from S onto $P(S)$. Form a subset T of S as follows: Let $s \in S$ be an element of T if and only if $s \notin f(s)$.) Prove that S has the same number of elements as a subset of $P(S)$.

4. Prove that the set of all finite subsets of the integers is countable, while the set of all subsets of the integers is uncountable.

5. A complex number z is said to be an *algebraic number* if it is the root of a polynomial equation in one indeterminate x with integer coefficients. Prove that the set of algebraic numbers is countable. Note that the rational numbers form a subset of the set of algebraic numbers.

1.4 ORDERED INTEGRAL DOMAINS

An *integral domain* is a set T such that given any two elements a and b of T we have unique elements $a + b$ and ab of T called the *sum* and *product* of a and b, respectively, which satisfy the following properties:*

D1) For any a, b, and c in T, $a + (b + c) = (a + b) + c$ and $a(bc) = (ab)c$. (These are the *associative laws* for addition and multiplication.)

D2) For any a and b in T, $a + b = b + a$ and $ab = ba$. (These are the *commutative laws*.)

D3) There is an element of T which we denote by 0 having the property that $a + 0 = a$ for any $a \in T$. (0 is the *identity* with respect to $+$.)

D4) Given any $a \in T$, the equation $a + x = 0$ has a unique solution which we denote by $-a$.†

D5) There is an element of T which we denote by 1 such that $a1 = a$ for any $a \in T$. (1 is the *identity* with respect to multiplication.)

D6) If a and b are elements of T such that $ab = 0$, then either $a = 0$ or $b = 0$.

D7) For any elements a, b, and c of T, $a(b + c) = ab + ac$. (That is, multiplication is *distributive* over addition.)

The integers, rational numbers, and real numbers are all examples of integral domains.

An integral domain T is said to be *ordered* if T contains a subset P, called the set of *positive* elements of T, having the following properties:

P1) For any $a \in T$, if $a \neq 0$, then either a or $-a$ is an element of P, but not both.

P2) If a and b are any elements of P, then $a + b$ and ab are both in P.

We then define $a < b$ if $b + (-a) \in P$ and $a \leqslant b$ if $a < b$ or $a = b$. The following proposition may now be proved.

Proposition 8: Let T be an ordered integral domain. Then

a) $0 < a$ if and only if $a \in P$.
b) If $a \leqslant b$, then $a + c \leqslant b + c$, a, b, and c elements of T;
c) If $a \leqslant b$ and $0 < c$, then $ac \leqslant bc$, a, b, and c elements of T;
d) $\leqslant$ defines a total ordering of T.

---

* A *binary operation* # on a set S is formally defined as a function from $S \times S$ into S; however, instead of writing $\#(s, s') = s''$, we usually write $s \# s' = s''$. Since the reader is almost certainly experienced with the use of many binary operations, we will not develop this formal approach. That is, it would probably only confuse matters to speak of addition and multiplication as functions.

† More formally, (D1) through (D4) tell us that an integral domain is a *commutative group* with respect to $+$ and a commutative *semigroup* with respect to multiplication.

We will not prove Proposition 8 or any of the algebraic consequences of (D1) through (D7). Our primary purpose in this section is to make precise some of the properties of the integers, properties which the reader has at least known implicitly since early schooling.

A *field* is an integral domain T of more than one element which satisfies the following additional property: Given any $a \in T$, $a \neq 0$, the equation $ax = 1$ has a unique solution in T which we denote by a^{-1}, or $1/a$.

The integers do not form a field, but the rational and real numbers each do.

An *ordered field* is a field which is also an ordered integral domain. The rational numbers and the real numbers each form ordered fields.

The set of integers can be characterized as an ordered integral domain whose set P of positive elements satisfies (N1) to (N3) (Section 1.2) with respect to the total ordering $\leqslant$ defined earlier in terms of P.

EXERCISES

1. Prove the following facts about any integral domain T; assume a, b, and c are elements of T.

a) $a + b = a + c$ implies $b = c$.
b) $(-a)(b) = -ab$.
c) $a0 = 0$.
d) $(-a)(-b) = ab$.
e) $ab = ac$ implies $b = c$, if $a \neq 0$.

2. Prove Proposition 8.

3. Let F be an ordered field. Prove:

a) $0 < a$ implies $0 < a^{-1}$.
b) $0 < a < b$ implies $0 < b^{-1} < a^{-1}$.
c) $0 \leqslant a^2$ for any $a \in F$.

4. Prove that no field containing only finitely many elements can be an ordered field.

2

THE REAL NUMBERS

2.1 THE INTEGERS AND THE RATIONAL NUMBERS

The purpose of this chapter is to "construct" the real numbers beginning with the set of integers. That is, we will basically accept the integers as given (although we will review the basic properties of the set of integers).* We will then show how to construct a set $\mathbb{Q}$ from the set $\mathbb{Z}$ of integers such that $\mathbb{Q}$ has all the properties we want and expect of the rational numbers. We will then use $\mathbb{Q}$ to construct the set $\mathbb{R}$ of real numbers.

THE INTEGERS

For the sake of review, uniformity, and precision of thought, we now state some essential properties of the set $\mathbb{Z}$ of integers.

1) Algebraically, $\mathbb{Z}$ forms an integral domain (cf. Section 1.4) with addition and multiplication as the operations.

2) The integers form a totally ordered set. The total ordering $\leqslant$ of $\mathbb{Z}$ can be characterized in terms of the subset P of positive elements of $\mathbb{Z}$. The subset P has the following properties:

    a) If n is any element of $\mathbb{Z}$, then one and only one of the following is true: $n = 0$, $n \in P$, or $-n \in P$ (in particular, $0 \notin P$).

    b) If n and m are elements of P, then $n + m$ and nm are also in P.

Given $n,\ m \in \mathbb{Z}$, we then define $n \leqslant m$ if $m - n \in P$ or $n - m = 0$. A *non-negative* element of $\mathbb{Z}$ is one which is either positive or 0; an element of $\mathbb{Z}$ which is not positive or 0 is said to be *negative*.

3) The integers have the further property that the set P of positive

---

elements satisfies properties (N1) through (N3) of Section 1.2 with respect to $\leqslant$.

Definition 1:   *An integral domain (possibly a field) R which has a subset P satisfying properties (a) and (b) under (2) is said to be an* **ordered integral domain.** *If R is also a field, then we call R an* **ordered field.**

If R is an ordered integral domain with total ordering $\leqslant$, and $a, b \in R$, then $a < b$ if $a \leqslant b$ but $a \neq b$. We may also denote $a \leqslant b$ by $b \geqslant a$ and $a < b$ by $b > a$.

Note that $<$ is not a partial ordering of R since for any $a \in R$, $a < a$. One can prove that $<$ is transitive and that given two *distinct* elements a and b of R, either $a < b$ or $b < a$, but not both.

The set $\mathbb{Z}$ of integers can be defined as an ordered integral domain in which the set of positive elements satisfies (N1) through (N3). Given any two ordered integral domains $\mathbb{Z}$ and $\mathbb{Z}'$ both of which have a set of positive elements satisfying (N1) through (N3), the elements, operations, and ordering of $\mathbb{Z}'$ can be labeled so that $\mathbb{Z}'$ is indistinguishable from $\mathbb{Z}$. (More formally, two such integral domains $\mathbb{Z}$ and $\mathbb{Z}'$ are isomorphic from an algebraic and order point of view.)

The following proposition summarizes some of the properties of $\mathbb{Z}$ which depend only on (1) and (2) and, hence, which pertain to any ordered integral domain. The proofs are fairly straightforward, and we leave most of them to the reader.

Proposition 1:   Let R be an ordered integral domain with set of positive elements P. Then

a) If $r \in R$, then $r \in P$ if and only if $0 < r$.
b) $\leqslant$ is a total ordering of R.
c) $r < s$ if and only if $s + (-r) \in P$.
d) If $r \leqslant s$, then $r + t \leqslant s + t$, for any $r, s, t \in R$.
e) If $r \leqslant s$ and $0 \leqslant t$, then $rt \leqslant st$, for any $r, s, t \in R$.
f) If $r \in R$, then r^2 is non-negative.

PROOF: a) If $r \in P$, then $r - 0 \in P$; hence $0 \leqslant r$. Since $r \in P$, $r \neq 0$ by (a) of (2), hence $0 < r$. If $0 < r$, then $0 \neq r$ and $r - 0 = r \in P$.

The following proposition gives certain properties of the integers which depend on the special properties of P and, for this reason, are not applicable to all ordered integral domains.

Proposition 2:   Let $\mathbb{Z}$ be the ordered integral domain of integers with set P of positive elements. Then

a) For any $z \in \mathbb{Z}$, there is no integer q such that $z < q < z + 1$.

b) (*First Principle of Finite Induction*) If some property S applies to 1 and if whenever S holds for some element n of P, S necessarily holds for $n + 1$, then S holds for each element of P.

The proofs of (a) and (b) are left to the reader.

THE RATIONAL NUMBERS

We normally think of a rational number as the quotient of two integers. Any rational number, though, can be represented in many ways as the quotient of integers. For example, $2/3 = 4/6 = 10/15 = -14/-21$, and so forth. Beginning with $\mathbb{Z}$, the ordered integral domain of integers, we now formally construct the rational numbers.

Let $\bar{\mathbb{Q}}$ be the subset of $\mathbb{Z} \times \mathbb{Z}$ consisting of all ordered pairs (m, n) of integers such that $n \neq 0$. (Informally, (m, n) represents m/n.) Given two such ordered pairs (m, n) and (p, q), we define (m, n) to be *equivalent* to (p, q) if $mq = np$; if (m, n) is equivalent to (p, q), we write $(m, n) \sim (p, q)$. (In the usual notation of the rational numbers, $m/n = p/q$ if and only if $mq = np$.) We define any element (m, n) of $\bar{\mathbb{Q}}$ together with all elements of $\bar{\mathbb{Q}}$ equivalent to (m, n) to be a *rational number*. That is, a rational number is an equivalence class of $\mathbb{Q}$. Denote the set of equivalence classes by $\mathbb{Q}$ and the equivalence class of (m, n) by m/n.

Given any two elements m/n and p/q of $\mathbb{Q}$, we set

$$(m/n) + (p/q) = (mq + np)/nq \tag{1}$$

$$(m/n)(p/q) = (mp/nq). \tag{2}$$

The operations defined by (1) and (2) are what we would expect on the basis of the way fractions are usually added and multiplied. Moreover, if we use any element of m/n (remember m/n is the equivalence class of (m, n)) to compute the sum in (1) or the product in (2), we still obtain $(mq + np)/nq$ or mp/nq, respectively. It follows then that addition and multiplication of rational numbers are well-defined; that is, depend only on the equivalence classes being added or multiplied and not on the representatives of the classes used to compute the sum or product. Although a rational number, strictly speaking, is a set of elements of $\mathbb{Q}$ (hence an element (m, n) of $\mathbb{Q}$ is merely a representative of a rational number m/n), since all representatives of the same rational number behave in essentially the same way as regards addition, multiplication, and the ordering of the rational numbers, we can use (m, n) to obtain the properties of m/n.

Observe for any $m/n \in \mathbb{Q}$, $0/1 + m/n = m/n + 0/1 = m/n$; hence $0/1$ is a zero element of $\mathbb{Q}$. Applying the definition of addition, we also find that $m/n + (-m/n) = 0/n = 0/1$; hence $-(m/n) = (-m)/n$. We can also identify each integer n with the element $n/1$ of $\mathbb{Q}$, and thus "embed" $\mathbb{Z}$ as a subset of $\mathbb{Q}$. (It is not hard to show then that the algebraic properties of $\mathbb{Z}$ are preserved by this "embedding." Once an ordering has been defined on $\mathbb{Q}$, we can prove the embedding also preserves the order properties of $\mathbb{Z}$.)

We now define a set P of positive elements for $\mathbb{Q}$. We let an element m/n of $\mathbb{Q}$ be in P if and only if mn is a positive integer.

Proposition 3:   Let $\mathbb{Q}$ and P be as defined. Then:

a) $\mathbb{Q}$ is an ordered field.
b) $0/1 < m/n$ if and only if $0/1 < n/m$.
c) If $0/1 < m/n < m'/n'$, then $0/1 < n'/m' < n/m$.
d) The rational numbers, being a subset of the countable set $\mathbb{Z} \times \mathbb{Z}$, form a countable set.

PROOF: We prove (b) and leave the other proofs to the reader. Since $0/1$ is the zero element of $\mathbb{Q}$ and $\mathbb{Q}$ is an ordered integral domain, $0/1 < m/n$ if and only if $m/n \in P$ by Proposition 1(a). But $m/n \in P$ if and only if mn is a positive integer and $n/m \in P$ if and only if nm is a positive integer. Since $mn = nm$, m/n and n/m are either both in P, and hence greater than $0/1$, or both outside of P.

We have outlined the construction of the field of rational numbers beginning with the ordered integral domain of integers. In the next section we will construct the field of real numbers from the rationals. Before doing this, however, we will show that the rationals are in a certain sense "incomplete"; the real numbers will be constructed by "completing" the rationals.

Example 1:   We first prove that given any positive rational number r such that $r^2 < 2$, there is a positive rational number s such that $r^2 < s^2 < 2$; also given any positive rational number r' such that $2 < r'^2$, there is a positive rational number s' such that $2 < s'^2 < r'^2$. Suppose that r is a positive rational number such that $r^2 < 2$. Then $(1/8)(2 - r^2)$ is positive; hence $r < r + (1/8)(2 - r^2)$. But $(r + (1/8)(2 - r^2))^2 < 2$. For

$$2 - (r + (1/8)(2 - r^2))^2 = (2 - r^2)(1 - r/4 - (2 - r^2)/64)$$

and $2 - r^2$ is positive, while $r/4 + (2 - r^2)/64 < 1$ (Exercise 9). Therefore $2 - (r + (1/8)(2 - r^2))^2$ is the product of two positive numbers and, hence, is positive.

If $2 < r'^2$ and r' is a positive rational, then $(1/8)(r'^2 - 2)$; hence setting $s' = r' - (1/8)(r'^2 - 2)$, we find $0 < s' < r'$, but $2 < s'^2 < r'^2$.

Let W be the set of rational numbers whose squares are less than 2. Then W has an upper bound, for example, 3. But the considerations of the preceding paragraphs indicate that if W is to have a least upper bound, then that least upper bound will have to be a number whose square is 2. We now prove that there is no rational number whose square is 2.

Suppose m/n is a rational number with $(m/n)^2 = 2$. We may assume that m and n are not both even; for if 2 divides both m and n, we can divide both m and n by a sufficiently high power of 2 to obtain a rational number $m'/n' = m/n$, where 2 does not divide both m' and n'. Now if 2 divides m^2, then 2 divides m. For $m = 2k$, or $m = 2k + 1$; if $m = 2k + 1$ for some integer k, then 2 does not divide $m^2 = 4k^2 + 4k + 1$. Similarly, if 2 divides n^2, then 2 divides n. Now $m^2/n^2 = 2$; hence $m^2 = 2n^2$. Therefore 2 does divide m^2; hence 2 divides m. But if 2 divides m, then $m^2 = 4k^2$ for some integer k. Consequently, $4k^2 = m^2 = 2n^2$, which implies that 2 divides n^2, and, hence, 2 divides n. This contradicts the assumption that 2 does not divide both m and n. Therefore $\sqrt{2}$ is not a rational number.

Not only have we shown that $\sqrt{2}$ is an "irrational" number, but we have produced a subset W of $\mathbb{Q}$ which has an upper bound, but no least upper bound. Moreover, the set W' of rational numbers greater than each element of W has a lower bound, but no greatest lower bound. There is a "gap" in the rational numbers between W and W' which will later be filled by $\sqrt{2}$.

EXERCISES

1. Prove (b) of Proposition 1.

2. Prove (c) through (f) of Proposition 1.

3. Prove Proposition 2. Use (a) to prove (b). To prove (c) assume that the set W of positive integers to which S does not apply is non-empty and let $n = \text{glb } W$. Show that n cannot be in W, contradicting the fact that W contains its glb.

4. Prove Proposition 3.

5. Prove that if m/n, m'/n', and p/q are rational numbers with $m/n = m'/n'$, then $m/n + p/q = m'/n' + p/q$ and $(m/n)(p/q) = (m'/n')(p'/q')$. Prove that $m/n < p/q$ if and only if $m'/n' < p/q$.

6. Prove the following about any ordered integral domain R.

a) If $r < s$ and $t < t'$, then $r + t < s + t'$ for any r, s, t, t' in R.

b) If $0 < t$ and $rt < st$, then $r < t$, for r, s, t in R.

c) If $0 < r < s$, then $r^2 < s^2$ for r, s in R.

d) There is no element s of R such that $s^2 = -1$.

7. a) Prove that for any $x \in \mathbb{Q}$, there is an integer n such that $n/1 > x$.

b) Prove that if p and q are positive rational numbers with $p < q$, then there is a least positive integer n such that
$$(n - 1)p \leqslant q < np.$$

8. Prove that there is no rational number q such that $q^2 = 3$.

9. Prove that if r is a positive rational number and $r^2 < 2$, then $r/4 + (2 - r^2)/64 < 1$.

2.2 DEDEKIND CUTS

We now have the ordered field $\mathbb{Q}$ of rational numbers. We will use $\mathbb{Q}$ to construct another ordered field which contains the rationals as a subfield, but which is "complete" in the sense that any non-empty subset of the new field which has an upper bound has a least upper bound. We begin with the following definition.

Definition 2:   *A subset A of the field $\mathbb{Q}$ of rational numbers is said to be a **Dedekind cut**,* or simply a **cut**, if*

a) *A is non-empty, but $A \neq \mathbb{Q}$.*

b) *If $q \in A$ and $q' < q$, then $q' \in A$.*

c) *No element of A is an upper bound of A. That is, A contains no element q such that $q' \leqslant q$ for each element q' of A.*

Small letters such as p, q, and r will be used to denote rational numbers, while A, B, C, and D will be used to represent cuts. Since it will turn out that cuts are in fact real numbers, this notation will have to be modified after the field of real numbers has been defined. We now prove certain basic properties of cuts.

Proposition 4:   If A is a cut and $q \notin A$, then q is an upper bound for A.

PROOF: If q is not an upper bound for A, then $q < p$ for some $p \in A$; therefore $q \in A$ by Definition 2(b).

---

* Richard Dedekind (1831–1916) was one of several mathematicians of the latter part of the 19th century who tried to put the real numbers and calculus on a solid foundation by deriving them from the natural numbers. It is his construction we use in this chapter.

Proposition 5:   If a subset A of the rational numbers consists of all rational numbers less than (but not equal to) some rational number p, then A is a cut and $p = \text{lub } A$.

PROOF: Clearly $A \neq \varnothing$, and since $p \notin A$, $A \neq \mathbb{Q}$. If $q \in A$ and $q' < q$, then $q' < p$; hence $q' \in A$. If $q \in A$, then $q < (1/2)(p + q) < p$; therefore A contains no upper bound of A. Consequently A is a cut. Since p is an upper bound of A and since $q < p$ implies $q \in A$, $p = \text{lub } A$.

Definition 3:   *If a cut A consists of all rational numbers less than some rational number p, then A is said to be a **rational cut**. A may then be denoted by $\bar{p}$.*

We shall see that not all cuts are rational.

Since cuts are subsets of the set of rational numbers, the set of cuts can be partially ordered by $\subseteq$. We use $A \subset B$ to denote $A \subseteq B$, but $A \neq B$. The next proposition proves that the set of cuts is in fact totally ordered.

Proposition 6:   If A and B are cuts and $A \nsubseteq B$, then $B \subset A$.

PROOF: Suppose $A \nsubseteq B$. Then there must be some rational number p which is an element of A, but with $p \notin B$. Assume that $B \nsubseteq A$. Then there must be $p' \in B \sim A$. Now $p \neq p'$; hence $p < p'$, or $p' < p$. In the latter case, $p' \in A$ since $p \in A$. In the former, $p \in B$ since $p' \in B$. In either case we have a contradiction; hence $B \subset A$.

Corollary: Given cuts A and B, $A \subset B$ if and only if there is a rational number $p \in B \sim A$.

PROOF: If $p \in B \sim A$, then $B \neq A$ and $B \nsubseteq A$; therefore $A \subset B$. If $A \subset B$, then there is some $p \in B \sim A$.

The set of cuts then forms a totally ordered set. We will define operations on the set of cuts which will make this set into an ordered field. We denote the set of cuts by $\mathbb{R}$.

Definition 4:   *If A and B are any cuts, define $A + B = C$, where C is the set of all rational numbers of the form $p + q$, where $p \in A$ and $q \in B$; that is, $A + B = C = \{p + q \mid p \in A \text{ and } q \in B\}$.*

If A is any cut, define $-A = \{q \in \mathbb{Q} \mid -q$ is an upper bound for A, but $-q$ is not the least upper bound for $A\}$.

*A cut A is said to be **positive** if $\bar{0} \subset A$. A is said to be **non-negative** if $\bar{0} \subseteq A$. If $A \subset \bar{0}$, then A is said to be **negative**.*

If A and B are any two non-negative cuts, define $AB = C$, where $C = \{r \in \mathbb{Q} \mid r$ is negative, or $r = pq$, where $p \in A$, $q \in B$, and both p and q are non-negative$\}$.

Before completing the definitions which will make $\mathbb{R}$, the set of cuts, into an ordered field, we must show that the sum and product of two cuts is again a cut and prove some of the basic properties of the addition and multiplication.

Lemma 1: Let $\{A_i\}$, $i \in I$, be any non-empty family of cuts. Then $\bigcup_I A_i$ is either $\mathbb{Q}$ or a cut.

PROOF: $A = \bigcup_I A_i$. Since $\{A_i\}$, $i \in I$, is non-empty and each A_i is non-empty, A is non-empty. Suppose $A \neq \mathbb{Q}$; we will prove that A is a cut.

Assume $q \in A$ and $q' < q$. Since $q \in A$, $q \in A_i$ for some $i \in I$. Since A_i is a cut, $q' \in A_i$; therefore $q' \in A$. Suppose now that A contains some element q such that $q' \leqslant q$ for each $q' \in A$. Since $q \in A$, $q \in A_i$ for some $i \in I$. Therefore $q' < q$ for each q' in A_i, a contradiction to the assumption that A_i is a cut. We have thus established that if $A \neq \mathbb{Q}$, then A is a cut.

Lemma 2: Let A be any cut and $b \in \mathbb{Q}$. Set $A + b = \{a + b \mid a \in A\}$ and $Ab = \{ab \mid a \in A\}$. Then $A + b$ is a cut and Ab is a cut when $0 < b$.

PROOF: We prove that $A + b$ is a cut and leave the proof that Ab is a cut when $b > 0$ as an exercise. Since A is non-empty, $A + b$ is non-empty. Because A is a cut, it is possible to find $q \in \mathbb{Q}$ such that $x < q$ for all $x \in A$. Then $q + b \notin A + b$; hence $A + b \neq \mathbb{Q}$. Suppose that $q \in A + b$ and $q' < q$. Since $q \in A + b$, $q = q_1 + b$ for some $q_1 \in A$. Because $q' < q$, $q' = q_2 + b$, where $q_2 < q_1$. But since A is a cut, $q_2 \in A$; therefore $q' \in A + b$. Assume now $A + b$ contains some q such that for each $q' \in A + b$, $q' \leqslant q$. Then for each $q'' \in A$, $q'' \leqslant q - b$, which contradicts the assumption that A is a cut. Therefore $A + b$ is a cut.

Proposition 7: If A and B are any cuts, $A + B$ is a cut.

PROOF: We have $A + B = \bigcup_{b \in B}(A + b)$. By Lemma 2, $A + b$ is a cut for each $b \in B$; therefore $A + B$ is a cut by Lemma 1 if $A + B \neq \mathbb{Q}$. Since A and B are both cuts, there are rational numbers q and q' such that $x < q$ for each $x \in A$ and $y < q'$ for each $y \in B$. Then $q + q' \notin A + B$; hence $A + B \neq \mathbb{Q}$.

Proposition 8: Let A be any cut and p be any positive rational number. Then there are rational numbers $s \in A$ and $t \notin A$ such that $t - s = p$ and t is not the least upper bound of A.

PROOF: Choose any $q \in A$. Set $b_n = q + np$, $n = 0, 1, 2, \ldots$. Using Exercise 7 of the previous section we see that there is some positive integer n such that $b_n \notin A$. Since the non-negative integers are well-ordered and $b_0 \in A$, there is a least integer m such that $b_{m-1} = q + (m - 1)p \in A$, but $b_m = q + mp \notin A$.

Case 1: $b_m = $ lub A. Then there exists a positive rational number e such that $q + (m - 1)p + e \in A$. Set $s = q + (m - 1)p + e$ and $t = b_m + e = q + mp + e$.

Case 2: $b_m \neq$ lub A. Set $t = b_m$ and $s = b_{m-1}$.

Proposition 9: $\mathbb{R}, +$ satisfies (D1) through (D4) with respect to $+$.

PROOF: We leave the proofs that $+$ is commutative and associative as exercises. These properties follow at once from the corresponding properties of addition of rational numbers.

$A + \bar{0} = A$ for any cut A: Suppose first that $p \in A$. Then there is an element q of A with $p < q$. Now $p = q + (p - q)$ and $p - q \in \bar{0}$ since $p - q < 0$; therefore $p \in A + \bar{0}$; hence $A \subseteq A + \bar{0}$. Suppose now that $p + q \in A + \bar{0}$ with $p \in A$ and $q \in \bar{0}$. Then q is negative; hence $p + q < p$, from which it follows that $p + q \in A$. Consequently, $A + \bar{0} \subseteq A$. Therefore $A = A + \bar{0}$.

$-A$ is a cut: This follows at once from Lemma 1 and the fact that

$$-A = \bigcup \{\bar{q} \mid q \in \mathbb{Q}, -q \text{ is an upper bound of } A\} \neq \mathbb{Q}.$$

$A + (-A) = \bar{0}$: If $p + q \in A + (-A)$, $p \in A$, and $q \in -A$, then $p < -q$; hence $p + q < q + (-q) = 0$. Therefore $p + q \in \bar{0}$; hence $A + (-A) \subseteq \bar{0}$. Suppose $p \in \bar{0}$. Then $p < 0$, and, hence, $0 < -p$. Then by Proposition 8, there are $s \in A$ and $t \notin A$ such that $t - s = -p$, where $t \neq $ lub A. Therefore $-t \in -A$. Since $p = s + (-t)$, $p \in A + (-A)$. Therefore $\bar{0} \subseteq A + (-A)$; hence $\bar{0} = A + (-A)$.

This completes the proof that $\mathbb{R}, +$ satisfies (D1) through (D4).

Proposition 10:   If A and B are non-negative cuts, then AB is a cut.

PROOF: Proposition 10 follows at once from Lemma 2 and the fact that

$$AB = \bigcup_{\substack{b \in B \\ 0 \leq b}} (Ab) \neq \mathbb{Q}.$$

We leave the proof that $AB \neq \mathbb{Q}$ as an exercise.

Proposition 11:   If A and B are positive cuts, then $A + B$ and AB are positive.

PROOF: Since A and B are positive, there are positive elements p and q of A and B, respectively. Therefore $p + q$ and pq are positive elements of $A + B$ and AB, respectively. Consequently, $p + q \in (A + B) \sim \bar{0}$ and $pq \in (AB) \sim \bar{0}$. Applying the corollary to Proposition 6, we obtain $\bar{0} \subset A + B$ and $\bar{0} \subset AB$; hence $A + B$ and AB are both positive.

Proposition 12:   If A is any cut, then one and only one of the following is true: $A = 0$, A is positive, or $-A$ is positive.

PROOF: Since A is positive if and only if $\bar{0} \subset A$, it is impossible that A be both $\bar{0}$ and positive. If A is negative, then there is some rational number $p < 0$ such that p is an upper bound for A, but $p \neq$ lub A. Then $-p \in -A$ and $0 < -p$; therefore $-A$ is positive. If $-A$ is positive, then 0 is an upper bound for A, but not the least upper bound; therefore A is negative. Hence A is negative if and only if $-A$ is positive. Proposition 12 follows at once.

Proposition 13: If A, B, and C are positive cuts, then $A(B + C) = AB + AC$.

PROOF: $A(B + C)$ consists of all the negative rational numbers together with all rational numbers of the form $r(s + t)$, $r \in A$, $s \in B$, and $t \in C$, r, s, and t non-negative. Since r, s, and t are non-negative, $r(s + t) = rs + rt$ is also an element of $AB + AC$. Therefore $A(B + C) \subseteq AB + AC$. AB and AC both contain all the negative rational numbers. Suppose $rs + r't$ is any element of $AB + AC$ such that $r, r' \in A$, $s \in B$, $t \in C$, and $r, r', s, t \geqslant 0$. Then either $0 \leqslant r \leqslant r'$ or $0 \leqslant r' \leqslant r$. In the former case $rs + r't \leqslant r's + r't \in A(B + C)$; in the latter case $rs + r't \leqslant rs + rt \in A(B + C)$. In either case $rs + r't \in A(B + C)$ since $A(B + C)$ is a cut. Therefore $AB + AC \subseteq A(B + C)$; hence $A(B + C) = AB + AC$.

In order to complete the proof that $\mathbb{R}$ is a field, we must show that the non-zero elements of $\mathbb{R}$ form a commutative group with respect to multiplication.* To this end we must define the multiplicative inverse of any non-zero element of $\mathbb{R}$. Moreover, since multiplication is only defined between non-negative cuts, we must extend the definition of multiplication to all of $\mathbb{R}$.

Once it is established that $\mathbb{R}$ is a field we want to show that $\mathbb{R}$ is an ordered field. We have two possible orderings for $\mathbb{R}$: the first obtained by ordering the cuts by $\subset$ and the second obtained by using the set of positive cuts. These two methods of ordering $\mathbb{R}$ are equivalent as follows easily from Propositions 11 and 12.

Once we have established that $\mathbb{R}$ is an ordered field, it still remains to be shown that $\mathbb{R}$ indeed "looks like" the field of real numbers. We close this section by extending our definition of multiplication and proving that $\mathbb{R}$ is a field.

Definition 5: *Let A be a positive cut. Define A^{-1} to consist of all non-positive rational numbers together with all rational numbers of the form $1/p$, where p is an upper bound of A, but not lub A.*

Proposition 14: A^{-1} as defined for any positive cut A is a cut.

We leave the proof of Proposition 14 to the reader.

---

* See Section 1.4.

Proposition 15:   The set P of positive cuts with multiplication as the operation satisfies the following for any A, B, $C \in P$: $AB = BA$, $A(BC) = (AB)C$, $A \cdot \bar{1} = A$, and $AA^{-1} = \bar{1}$.

PROOF: The proofs that $AB = BA$, $A(BC) = (AB)C$, and $A\bar{1} = A$ are left as exercises. It remains to be shown that $AA^{-1} = \bar{1}$ for any positive cut A. AA^{-1} contains all the non-positive rational numbers. Suppose $q(1/p) \in AA^{-1}$, $q \in A$ and $1/p \in A^{-1}$, q and p positive. Then $0 < q < p$; hence $0 < 1/p < 1/q$. Therefore $q(1/p) < q(1/q) = 1$; hence $q(1/p) \in \bar{1}$. Therefore $AA^{-1} \subseteq \bar{1}$. Now $\bar{1}$ also contains all the non-positive rational numbers. Suppose s is any positive element of $\bar{1}$. Then there is an integer M such that M/s is not an element of A, but $s^2/M \in A$ (see Exercise 1). Then $s/M \in A^{-1}$ and $(s^2/M)(M/s) = s \in AA^{-1}$. Consequently, $\bar{1} \subseteq AA^{-1}$; hence $AA^{-1} = \bar{1}$.

We now extend multiplication to all of $\mathbb{R}$.

Definition 6:   *If A and B are non-negative cuts, then AB has already been defined. If A is negative and B is non-negative, define $AB = -((-A)B)$. If A is non-negative and B is negative, define $AB = -(A(-B))$. If both A and B are negative, define $AB = (-A)(-B)$.*

Proposition 16:   The set $\mathbb{R}$ with multiplication and addition as defined previously and the ordering given by the set of positive cuts is an ordered field. Moreover, given cuts A and B, $A + (-B)$ is positive if and only if $B \subset A$; hence the ordering from the positive elements is the same as the ordering of $\mathbb{R}$ given by $\subset$.

The details of the proof not already supplied in previous proofs are left as exercises.

EXERCISES

1. Let A be any positive cut and p be any positive rational number.

a) Let q be any rational number. Prove that there is some positive integer n such that $q + np \notin A$.

b) Prove that there is some positive integer M such that $M/p \notin A$, but $p^2/M \in A$.

2. Prove Proposition 14.

3. Prove those parts of Proposition 15 whose proofs were left as exercises.

4. Prove those parts of Proposition 16 not proved in previous propositions.

5. Prove that addition of cuts is commutative and associative.

6. Prove the following statements directly without appeal to the fact that $\mathbb{R}$ is an ordered field.

    a) If A is negative, then A^2 is positive.
    b) If $A \subset B \subset \bar{0}$, then $B^{-1} \subset A^{-1} \subset \bar{0}$.
    c) If $\bar{0} \subset A \subset \bar{1}$, then $\bar{1} \subset A^{-1}$.

7. Let p and q be any rational numbers. Prove that $\bar{p} + \bar{q} = \overline{p + q}$ and $\bar{p}\bar{q} = \overline{pq}$.

8. Let A be a cut and $b > 0$. Prove that Ab is a cut.

9. Prove that if A and B are cuts, then $AB \neq \mathbb{Q}$.

2.3 BASIC PROPERTIES OF THE REAL NUMBERS

Definition 7: *We define the ordered field* $\mathbb{R}$ *of cuts to be the* **field of real numbers** *and each element of* $\mathbb{R}$ *to be a* **real number.** *A rational cut will be called a* **rational real number,** *or simply a* **rational number.** *Any real number which is not rational will be said to be* **irrational.**

The ordered field $\mathbb{Q}$ of rational numbers as defined in Section 2.1 can be considered to be an ordered subfield of $\mathbb{R}$ by identifying $p \in \mathbb{Q}$ with $\bar{p} \in \mathbb{R}$. (The correspondence $p \to \bar{p}$ is one-one, onto the set of rational cuts, order-preserving, and a *field isomorphism;* that is, it is an embedding of $\mathbb{Q}$ into $\mathbb{R}$ from both an algebraic and order point of view. Despite the somewhat technical nature of the statement of this basic relationship between $\mathbb{Q}$ and $\mathbb{R}$ and the numerous details that must be supplied for a formal proof, the truth of the statement should be fairly obvious to the reader. The reader should supply proofs for any portions of the claims that he is unsure of.)

Proposition 17 (*Denseness of the Rational Numbers in the Real Numbers*): If A and B are real numbers such that $A \subset B$, then there is a rational number $\bar{p}$ such that $A \subset \bar{p} \subset B$. (Informally, between any two real numbers there is a rational number.)

PROOF: If $A \subset B$, then there is a rational number $p \in B \sim A$. Therefore $A \subset \bar{p} \subset B$.

Proposition 18:   If A is any real number and p is any rational number, then $p \in A$ if and only if $\bar{p} \subset A$.

Proof: If $\bar{p} \subset A$, then there is $t \in \mathbb{Q}$ such that $t \in A \sim \bar{p}$. Therefore $p < t$ and $t \in A$; hence $p \in A$. If $p \in A$, then $\bar{p} \subseteq A$. But $p \notin \bar{p}$; hence $\bar{p} \subset A$.

Despite the lengths to which we have gone to define the real numbers, we still do not know that we have anything essentially different from $\mathbb{Q}$. In other words, it has not been shown that there is such a thing as an irrational number; perhaps every cut is a rational cut. We must also verify in some way that the "gaps" that exist in $\mathbb{Q}$ have been "filled in" in $\mathbb{R}$. To this latter end we prove the following proposition.

Proposition 19:   Suppose S and T are non-empty subsets of $\mathbb{R}$ such that $\mathbb{R} = S \cup T, S \cap T = \varnothing$, and each element of T is an upper bound for S (that is, $A \in T$ implies $B \subset A$ for each $B \in S$). Then there is a real number U such that $U = \text{lub } S = \text{glb } T$. (But another way, either S contains a greatest element, or T contains a least element.)

Proof: Set $U = \bigcup_{A \in S} A$. Then by Lemma 1, U is a cut, and, hence, is a real number. We leave it to the reader to prove that $U = \text{lub } S = \text{glb } T$.

Proposition 20:   Let W be any non-empty subset of $\mathbb{R}$ which has an upper bound. Then W has a least upper bound.

Proof: Let S be the set of real numbers A such that $A \subseteq C$ for some element C of W. Set $T = \mathbb{R} \sim S$. Since W has an upper bound, $T \neq \varnothing$. Since $W \neq \varnothing$, $S \neq \varnothing$. Each element of T is an upper bound for W and $\mathbb{R} = S \cup T$. If B is any element of T and A is any element of S, then $A \subseteq C$ for some element C of W and $C \subset B$; hence $A \subset B$. The hypotheses of the previous proposition are satisfied by S and T, therefore there is a real number U such that $U = \text{lub } S$. The proof that $U = \text{lub } W$ is left as an exercise.

Corollary: Let W be any non-empty subset of $\mathbb{R}$ which has a lower bound. Then W has a greatest lower bound.
The proof is left as an exercise.

If $\mathbb{R}$ consisted solely of rational numbers, then there would be non-empty subsets of $\mathbb{R}$ which have upper bounds but no least upper bounds (just as there were in $\mathbb{Q}$). Proposition 20 then assures of the existence of irrational numbers. Consequently, we have the following definition.

Definition 8:   *Let W be a subset of the set of real numbers. We say that W is **bounded above** if W has an upper bound; W is **bounded below** if W has a lower bound. We say that W is a **bounded set** if W is bounded above and below.*

From Proposition 20 and its corollary, we have at once that any non-empty bounded set has both a least upper bound and greatest lower bound. The next example illustrates properties of the lub and glb.

Example 2:   Let W be any non-empty bounded set of real numbers. For any $B \in \mathbb{R}$, define $W + B = \{A + B \mid A \in W\}$. We now show that lub $(W + B) =$ lub $W + B$. Since $A \subseteq$ lub W for each $A \in W$, we have $A + B \subseteq$ lub $W + B$ (since $\mathbb{R}$ is an ordered field); hence lub $W + B$ is an upper bound for $W + B$. If lub $W + B$ is not lub $(W + B)$, then there is an upper bound V of $W + B$ with

$$A + B \subseteq V \subset \text{lub } W + B, \quad \text{for all } A \in W.$$

From this it follows then that $A \subseteq V + (-B) \subset$ lub W for $A \in W$, which contradicts lub W as the least upper bound of W. Therefore lub $W + B =$ lub $(W + B)$.

We can also prove that glb $(W + B) =$ glb $W + B$.

Proposition 21:   The set of irrational numbers is non-empty.

We have already shown that between any two real numbers there is a rational number. We will prove that between every two real numbers there is also an irrational number. Preliminary to proving this, we prove two propositions.

Proposition 22:   If A is irrational and $\bar{p}$ is rational, then $A + \bar{p}$ is irrational and $A\bar{p}$ is irrational if $\bar{p} \neq \bar{0}$.

PROOF: If $A + \bar{p} = \bar{q}$, q rational, then $A = \bar{q} - \bar{p}$ which is rational. If $A\bar{p} = \bar{q}$, then $A = \bar{q}(\bar{p})^{-1} = \bar{q}(\overline{p^{-1}})$, which is again rational.

Proposition 23 (*The Archimedean Property of the Real Numbers*):   For any real number $A \supset \bar{0}$, there is a non-negative integer n such that $A \subset \bar{n}$.

PROOF: Let $W = \{\bar{n} \mid n \text{ a positive integer}\}$. If no $\bar{n} \in W$ is greater than A, then A is an upper bound for W; hence W has a least upper bound U. There is, however, some rational number p such that $U \subset \bar{p}$. This, in turn, implies that p is greater than any positive integer n, a contradiction. Therefore Proposition 23 must hold.

Corollary: Given any real numbers A and B with $A \subset B$ and any positive real number C, there is a non-negative integer n such that

$$A + \bar{n}C \subset B \subset A + (\overline{n + 1})C.$$

We leave the proof of this corollary to the reader.

Proposition 24:   Given any rational numbers p and q such that $p < q$, there is an irrational number A such that $\bar{p} \subset A \subset \bar{q}$.

PROOF: There is at least one irrational number B. There is an integer m such that $\bar{p} \subset B + \bar{m}$; moreover, $B + \bar{m}$ is an irrational number. Let $A' = B + \bar{m}$. Since $p < q$, $\bar{q} - \bar{p}$ is a positive real number. There is therefore a non-negative integer n such that $\bar{p} + \bar{n}(\bar{q} - \bar{p}) \subset A' \subset \bar{p} + \overline{(n + 1)}(\bar{q} - \bar{p})$. We leave it to the reader to show that if we set $A = A' - \bar{n}(\bar{q} - \bar{p})$, then A is irrational and $\bar{p} \subset A \subset \bar{q}$.

Corollary: If A and B are any real numbers such that $A \subset B$, then there is an irrational number C such that $A \subset C \subset B$.

PROOF: By a double application of Proposition 17 there are rational numbers p and q such that $A \subset \bar{p} \subset \bar{q} \subset B$. By Proposition 24 then there is an irrational number C such that $A \subset \bar{p} \subset C \subset \bar{q} \subset B$.

Now that we have defined the field $\mathbb{R}$ of real numbers and rigorously proved some of the basic properties of $\mathbb{R}$, we will revert to the more familiar notation for the real numbers; that is, p, q, r, and so forth will represent real numbers, either rational or irrational, and $<$ will be used in place of $\subseteq$ to indicate the ordering. We will assume that the integers and the rational numbers are simply subsets of $\mathbb{R}$ and make no distinction between p and $\bar{p}$.

EXERCISES

1. Prove that U as defined in the proof of Proposition 19 is both lub S and glb T.

2. Prove that the U in the proof of Proposition 20 is lub W.

3. Prove that A as defined in the proof of Proposition 24 has the properties claimed for it.

4. Prove that there is a real number r such that $r^2 = 2$; by what was proved in Example 1, r must be irrational.

5. Prove that there is a real number r such that $r^2 = 3$.

6. Prove the corollary to Proposition 20. Hint: Consider the set $-W = \{-w \mid w \in W\}$ and look at its lub.

7. Prove the corollary to Proposition 20 directly without appeal to Proposition 20. Prove that Proposition 20 is true if and only if its corollary is true.

8. Prove the corollary to Proposition 23.

9. Suppose W is a bounded subset of $\mathbb{R}$, B is a cut, and $W + B$ is defined as in Example 2. Prove that $\mathrm{glb}(W + B) = \mathrm{glb}\, W + B$. Suppose $B \supset \bar{0}$. Define $BW = \{BA \mid A \in W\}$. Prove that $\mathrm{glb}\, BW = B\, \mathrm{glb}\, W$.

10. Let a and b be positive real numbers. Prove $a^n < b^n$ if and only if $a < b$.

3

SOME BASIC TOPOLOGY
OF THE REAL NUMBERS

3.1 THE ABSOLUTE VALUE METRIC. INTERVALS

For the remainder of this text, as before $\mathbb{N}$, $\mathbb{Z}$, $\mathbb{Q}$, and $\mathbb{R}$ will denote the sets of positive integers, integers, rational numbers and real numbers, respectively.

THE ABSOLUTE VALUE

In his earlier experience with real numbers, the reader was almost certainly told that the "distance" between two real numbers a and b is given by

$$|a - b|, \tag{1}$$

the *absolute value* of the difference of a and b.* The absolute value of a real number was defined in the following way.

Definition 1: *Let r be any real number. Then $|r|$, the **absolute value** of r, is defined by:*

$$|r| = \begin{cases} r & \text{if } r \geqslant 0, \\ -r & \text{if } r < 0. \end{cases}$$

---

* The reader was also taught that the real numbers could be represented as points of a straight line. We will use this representation in illustrations.

*Given two real numbers a and b, the **absolute value distance,** or simply the **distance,** between a and b, is given by* (1).

Some of the basic properties of the absolute value and distance are summarized in the following proposition. The proof of this proposition is but a straightforward application of Definition 1. We leave the reader to supply for himself such details as he may not have encountered in his study of calculus.

Proposition 1

a) For any $r \in \mathbb{R}$, $|r| \geqslant 0$; hence the distance between any two real numbers is a non-negative real number.

b) For $r \in \mathbb{R}$, $|r| = 0$ if and only if $r = 0$; hence given a and b in $\mathbb{R}$, the distance from a to b is 0 if and only if $a = b$.

c) For any $r \in \mathbb{R}$, $|r| = |-r|$; hence given a and b in $\mathbb{R}$, $|a - b| = |b - a|$ (that is, the distance from a to b is the same as the distance from b to a.

d) (*The Triangle Inequality*) Given real numbers r and r', $|r + r'| \leqslant |r| + |r'|$; hence given a, b, and c in $\mathbb{R}$, $|a - c| \leqslant |a - b| + |b - c|$ (that is, the distance from a to b plus the distance from b to c is at least as great as the distance from a to c).

e) Given a and b in $\mathbb{R}$, $\left||a| - |b|\right| \leqslant |a - b|$ (that is, the distance between a and b is at least as large as the distance between $|a|$ and $|b|$).

The absolute value distance defined for the set $\mathbb{R}$ of real numbers is but a special case of the general notion of *metric* on a set. A metric is a measure of distance; its definition is as follows.

Definition 2: *Let X be any set. A function $D: X \times X \to \mathbb{R}$ is said to be a **metric,** or **distance function,** on X if D satisfies the following properties:*

a) $D(x, y) \geqslant 0$ for any $x, y \in X$.

b) $D(x, y) = 0$ if and only if $x = y$.

c) $D(x, y) = D(y, x)$ for any $x, y \in X$.

d) $D(x, y) + D(y, z) \geqslant D(x, z)$, for any $x, y, z \in X$.

The set X with metric D is said to be a *metric space*. We may denote this metric space by X, D.

Compare the properties of D listed in (a) to (d) of Definition 2 with the corresponding properties of the absolute value distance given in Proposition 1. The absolute value distance defined for the set of real numbers is but one metric which might be defined on $\mathbb{R}$. Nevertheless, the absolute value metric is a particularly important metric because it is very closely tied to the order properties of $\mathbb{R}$. Preparatory to discussing the relationship between the order and metric properties of $\mathbb{R}$, we make several definitions.

INTERVALS AND OPEN SETS

Definition 3:   *Let a and b be two real numbers such that $a < b$. Set*

$$(a, b) = \{x \in \mathbb{R} \mid a < x < b\}, \tag{2}$$

$$[a, b] = \{x \in \mathbb{R} \mid a \leqslant x \leqslant b\}, \tag{3}$$

$$[a, b) = \{x \in \mathbb{R} \mid a \leqslant x < b\}, \quad \text{and} \quad (a, b] = \{x \in \mathbb{R} \mid a < x \leqslant b\}. \tag{4}$$

*We call (a, b) the **open interval** with **end points** a and b, while $[a, b]$ is the **closed interval** with end points a and b. The sets in (4) are said to be **half-open intervals**. (Note that a parenthesis indicates the omission of an end point, whereas square bracket indicates its inclusion.)*

*We say that a subset U of $\mathbb{R}$ is **order open** if given any point $u \in U$, there is an open interval which contains u and is contained in U.*

*Given any $a \in \mathbb{R}$ and any $p > 0$, we define the **p-neighborhood** of a to be $\{x \in \mathbb{R} \mid |a - x| < p\}$. We denote the p-neighborhood of a by $N(a, p)$. (Thus, $N(a, p)$ consists of all real numbers within distance p of a.)*

*A subset u of $\mathbb{R}$ is said to be **open** if given any $a \in U$, there is $p > 0$ such that $N(a, p) \subseteq U$.*

*A subset F of $\mathbb{R}$ is said to be **closed** if $\mathbb{R} \sim F$ is open.*

We now investigate the relationship between *order open* and *open*.

Proposition 2:   For any $a \in \mathbb{R}$ and $p > 0$, $N(a, p) = (a - p, a + p)$.

PROOF: If $x \in N(a, p)$, then $|a - x| < p$; hence it follows that $a - p < x < a + p$, or $x \in (a - p, a + p)$. On the other hand, if $x \in (a - p, a + p)$, then $a - p < x < a + p$; from which we have $-p < a - x < p$, or $|a - x| < p$. Therefore $x \in N(a, p)$. Consequently, $N(a, p) = (a - p, a + p)$.

Not only is every *p*-neighborhood an open interval, but every open interval is a *p*-neighborhood.

Proposition 3:   Let (a, b) be any open interval. Then

$$(a, b) = N\left(\frac{a + b}{2}, \frac{b - a}{2}\right).$$

The proof of Proposition 3 is left as an exercise (cf. Fig. 1).

The following example illustrates some of the properties of open intervals.

Example 1: We first note that any open interval is the union of closed intervals. For if $a < b$, then

$$(a, b) = \bigcup \{[a + p, b - p] \mid p > 0, p < (b - a)/2\} \tag{5}$$

We prove (5) as follows: Since $a < a + p < b - p < b$ for $0 < p < (b - a)/2$, we have each $[a + p, b - p] \subseteq (a, b)$ (Fig. 1); hence the right

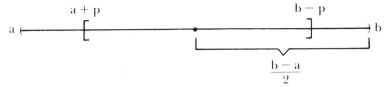

Figure 1

side of (5) is a subset of (a, b). If $x \in (a, b)$, then set p' equal to the minimum of $x - a$ and $b - x$ and select $0 < p < p'$. Then $x \in [a + p, b - p] \subseteq (a, b)$; hence (a, b) is also a subset of the right side of (5). Therefore (5) holds.

It should be noted too that for any x in an open interval (a, b), the maximum $p > 0$ such that $N(x, p) \subseteq (a, b)$ is the minimum of $x - a$ and $b - x$ (Fig. 2). Thus p depends on x. Moreover, if some $p > 0$ is such that

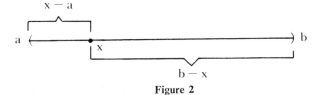

Figure 2

$N(x, p) \subseteq (a, b)$ and $0 < p' < p$, then $N(x, p')$ is also a subset of (a, b).

Since every p-neighborhood is an open interval and every open interval is a p-neighborhood, the set of p-neighborhoods is the same as the set of open intervals. Moreover, we have the following proposition.

Proposition 4: A subset U of $\mathbb{R}$ is order open if and only if U is open.

PROOF: Suppose U is an order open set and $u \in U$. Then there is an open interval (a, b) such that $u \in (a, b) \subseteq U$. Let p be the lesser of $u - a$ and $b - u$. Then $N(u, p) \subseteq U$; hence U is also an open set. On the other hand, if U is an open set, then it follows at once from the fact that any p-neighborhood of any point is an open interval containing that point that U is also order open. Therefore a set U is open if and only if it is order open.

The next proposition gives some of the fundamental properties of open sets.

Proposition 5

a) $\mathbb{R}$ and $\varnothing$ are open sets.

b) The intersection of any two open sets is again an open set.

c) The union of any family of open sets is again an open set.

d) Any p-neighborhood, and hence also any open interval, is an open set.

PROOF

(a) Given any $a \in \mathbb{R}$, $N(a, p) \subseteq \mathbb{R}$ for any $p > 0$. The same is true for any $a \in \varnothing$ since there is no $a \in \varnothing$.

(b) Let U and V be open sets such that $U \cap V \neq \varnothing$ and suppose $a \in U \cap V$. Since $a \in U$ and U is open, there is $p_1 > 0$ such that $N(a, p_1) \subseteq U$. Similarly, there is $p_2 > 0$ such that $N(a, p_2) \subseteq V$. Let p be the lesser of p_1 and p_2 . Then $N(a, p) \subseteq U \cap V$.

(c) Let $\{U_i\}$, $i \in I$, be any family of open sets and $a \in \bigcup_I U_i$. Then $a \in U_i$ for some $i \in I$. Since U_i is open, there is $p > 0$ such that $N(a, p) \subseteq U_i$. But then $N(a, p) \subseteq \bigcup_I U_i$. Therefore $\bigcup_I U_i$ is open.

(d) Let $N(a, p)$ be a p-neighborhood of a and suppose $b \in N(a, p)$. Set $q = p - |b - a|$. We now show that $N(b, q) \subseteq N(a, p)$. Suppose $x \in N(b, q)$. Then $|b - x| < q$. Therefore $|a - x| \leqslant |a - b| + |b - x| < |a - b| + (p - |a - b|) = p$; therefore $N(b, q) \subseteq N(a, p)$. Consequently, $N(a, p)$ is open.

Since the family of open subsets of $\mathbb{R}$ has the properties (a) through (c) of Proposition 5, we say that the open sets of $\mathbb{R}$ form a *topology* for $\mathbb{R}$. Any property of $\mathbb{R}$ which can be characterized in terms of this topology is said to be a *topological* property.* Because of Proposition 4, the topological properties of $\mathbb{R}$ can be studied using the ordering of $\mathbb{R}$, or the absolute value metric.

The topology of $\mathbb{R}$, that is, the family of open subsets of $\mathbb{R}$, is completely determined by the p-neighborhoods in the following manner.

Proposition 6:   A subset U of $\mathbb{R}$ is open if and only if U is the union of p-neighborhoods.

PROOF: If U is the union of p-neighborhoods, then U is the union of a family of open sets; hence U is open by Proposition 5(c). Suppose now that

---

* A *topology* $\mathscr{T}$ on a set X is a collection of subsets of X which satisfies (1) X and $\varnothing$ are members of $\mathscr{T}$, (2) the intersection of any two members of $\mathscr{T}$ is again a member of $\mathscr{T}$, and (3) the union of any family of members of $\mathscr{T}$ is a member of $\mathscr{T}$ (compare with (a) through (c) of Proposition 5). The members of $\mathscr{T}$ are said to be *open* sets. The study of the properties of topologies today is a very important branch of mathematics in its own right. One of the earliest topologies mathematicians studied (even before the term *topology* was invented) is the topology of open subsets of the real numbers. The properties of this topology have motivated much of the work done in modern point-set topology.

U is any open set. Then for each $x \in U$ we can find $p_x > 0$ such that $N(x, p_x) \subseteq U$. But then $U = \bigcup \{N(x, p_x) \mid x \in U\}$; consequently, U is the union of p-neighborhoods.

The following example shows that there are non-open and non-closed sets.

Example 2: Consider the half-open interval $[0, 1)$. This set is not open since there is no open interval which contains 0 and lies entirely in $[0, 1)$ (for any open interval which contains 0 must also contain numbers less than 0). On the other hand, this set is not closed since $\mathbb{R} \sim [0, 1) = \{x \in \mathbb{R} \mid x < 0 \text{ or } 1 \leqslant x\}$. Since there is no open interval which contains 1 and lies entirely in $\mathbb{R} \sim [0, 1)$, $\mathbb{R} \sim [0, 1)$ is not open; hence $[0, 1)$ is not closed.

The reader should prove that the closed interval $[0, 1]$ (indeed any closed interval) is a closed set. The only sets which are both open and closed are $\mathbb{R}$ and $\varnothing$ although we will not prove that these are the only closed *and* open sets in $\mathbb{R}$ until some time later.

The next proposition gives some fundamental properties of closed sets.

Proposition 7

a) $\mathbb{R}$ and $\varnothing$ are closed sets.
b) The union of any two closed sets is closed.
c) The intersection of any family of closed sets is closed.
d) Any one point subset is closed, as is any closed interval.

PROOF: We leave the proofs of (a) and (d) to the reader. Statements (b) and (c) are proved using the *DeMorgan formulas* as follows.

(b) Let F and F' be any two closed sets. Then $\mathbb{R} \sim F$ and $\mathbb{R} \sim F'$ are open. Now $\mathbb{R} \sim (F \cup F') = (\mathbb{R} \sim F) \cap (\mathbb{R} \sim F')$, which is the intersection of two open sets, and hence is open. But then $F \cup F'$ is closed.

(c) Let $\{F_i\}$, $i \in I$, be any family of closed sets. Then $\mathbb{R} \sim F_i$ is open for each $i \in I$. Now $\mathbb{R} \sim \bigcap_I F_i = \bigcup_I (\mathbb{R} \sim F_i)$, which is open since it is the union of open sets. Therefore $\bigcap_I F_i$ is closed.

Two important types of open sets will now be defined.

Definition 4: *For any real number a, set*

$$(a, \infty) = \{x \mid a < x\},$$

ánd

$$(-\infty, a) = \{x \mid x < a\}.$$

*These sets are called the **upper** and **lower open half-lines** with **end point** a.*

We leave it to the reader to show that any open half-line is an open set. We also leave it to the reader to formulate the definition of a closed half-line with end point a and to prove that any such closed half-line is closed.

We conclude this section by proving an important "separation" property of the real numbers which will later be used to prove that the limit of a convergent sequence is unique.

Proposition 8:   Given any two distinct real numbers a and b, there is $p > 0$ such that $N(a, p) \cap N(b, p) = \varnothing$.   (Informally, any two real numbers can be separated by means of disjoint neighborhoods (Fig. 3).)

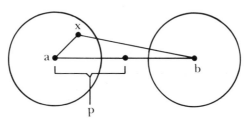

Figure 3

PROOF: Let $p = (1/2) |a - b|$. Suppose $x \in N(a, p) \cap N(b, p)$. Then $|a - x| < p$ and $|b - x| < p$. But $|a - b| \leqslant |a - x| + |x - b| = |a - x| + |b - x| < p + p = |a - b|$, a contradiction. Therefore

$$N(a, p) \cap N(b, p) = \varnothing.$$

EXERCISES

1. Prove (c), (d), and (e) of Proposition 1. The proofs are direct, but a number of different cases may have to be considered. For example, in the proof of (d), there are four cases: $r < 0$ and $r' < 0$; $r < 0$ and $r' \geqslant 0$, and so forth. Hint: To prove (e), $|a| = |a + (b - a)| \leqslant |a| + |b - a|$ or $|a| - |b| \leqslant |a - b|$. Similarly, $|b| - |a| \leqslant |a - b|$.

2. Prove that any closed interval is a closed set.

3. Prove that any open half-line is open. Prove that any closed half-line is closed.

4. Prove (a) and (d) of Proposition 7.

5. Given real numbers a, b, and c, prove that $a \leqslant b \leqslant c$ if and only if $|a - b| + |b - c| = |a - c|$.

6. Given $a, b \in \mathbb{R}$, define

$$D(a, b) = 3 \, |a - b|.$$

Prove that D is a metric on $\mathbb{R}$. Define a D-p-neighborhood and a D-open set in a manner similar to the way p-neighborhoods and open sets were defined relative to the absolute value metric. (In point of fact, many of the statements made about the real numbers in this section apply, possibly with an appropriate change of terminology, to any metric space.) Prove that a subset U of $\mathbb{R}$ is D-open if and only if it is open.

7. Restate and prove Proposition 7 for an arbitrary metric space.

8. For any a and b in $\mathbb{R}$, set

$$D'(a, b) = \begin{cases} 0 & \text{if} \quad a = b, \\ 1 & \text{if} \quad a \neq b. \end{cases}$$

Prove that D' is a metric on $\mathbb{R}$. Define D'-open sets in the appropriate way. Prove that every subset of $\mathbb{R}$ is D'-open; hence D' does not give the same topology of open sets that the absolute value metric gives.

9. Prove that every open subset of $\mathbb{R}$ contains infinitely many rational numbers. Topologically, this means that the rational numbers form a *dense* subset of $\mathbb{R}$.

3.2 CAUCHY SEQUENCES. METRIC COMPLETENESS

A notion that recurs frequently in analysis is that of *sequence*. In fact, the real numbers can be developed from the natural numbers using sequences, although we have used another approach—Dedekind cuts—in this text.

SEQUENCES

Definition 5: *A **sequence** of real numbers is a function from the set $\mathbb{N}$ of positive integers into $\mathbb{R}$. If s is a sequence, we generally denote $s(n)$ by s_n and s itself by either $\{s_n\}$, $n \in \mathbb{N}$, or*

$$s_1, s_2, s_3, \ldots.$$

A sequence s is said to be in a set W if $s_n \in W$ for all $n \in \mathbb{N}$.

Example 3: Define $s(n) = 1/n$ for each positive integer n. Then s is a sequence. This sequence may also be denoted by $\{1/n\}$, $n \in \mathbb{N}$, or 1, 1/2, 1/3, 1/4, ...

Example 4: The sequence

$$1, 2, 3, 4, \ldots$$

is defined by $s(n) = n$. It may also be denoted by $\{n\}$, $n \in \mathbb{N}$.

Although a sequence $\{s_n\}$, $n \in \mathbb{N}$, has infinitely terms in the sense that there is one term for each $n \in \mathbb{N}$, there do not have to be infinitely many different values for the s_n. This point is illustrated in the next example.

Example 5: Consider the sequence defined by $s_n = 1$. Each term of this sequence has the same value, 1. Though the sequence contains infinitely many terms, the set of values that the terms of the sequence assume is clearly finite.

Note that the ordering of the positive integers gives an ordering to the terms of a sequence. One can speak of the first term of a sequence, or the 10th term. Since $\mathbb{N}$ has no last element, a sequence has no last term.

The terms of the sequence given in Example 3 get closer and closer to 0, that is, approximate 0 with greater and greater accuracy, as we go further out in the sequence; in particular, it appears that while no term of the sequence is 0, we can approximate 0 by the terms of the sequence to any desired degree of accuracy provided only that we take s_n for sufficiently large n. The terms of the sequence in Example 4 get progressively larger, but they do not seem to approach any real number as a "limit"; that is, the terms of this sequence do not get and stay close to any real number, in other words, they do not tend to approximate any real number. In Example 5, each term of the sequence is a perfect approximation of 1.

The absolute value metric gives us a measure of how close two real numbers are to one another. With this in mind, we can formally state what it means for a sequence s to have a limit L.

Definition 6: *Let s be any sequence (of real numbers). We say that $s* **converges** *to, or has* **limit**, *L if given any positive number p, there is an integer M such that $n > M$ implies $s_n \in N(L, p)$. If s has limit L, we may write $s_n \to L$.*

Definition 6 simply states that s has the limit L if given any $p > 0$, s_n is within distance p of L provided that n is large enough.

Example 6: Let s be the sequence of Example 3. Then $s_n \to 0$. For let $p > 0$ and M be any integer greater than $1/p$. (Note that we are certain such an integer exists because of the Archimedean property of the real numbers.) Then $1/p > M$; hence if $n > M$, $|s_n - 0| = s_n = 1/n < 1/M < p$. That is, $n > M$ implies $s_n \in N(0, p)$. Since p was arbitrary, $s_n \to 0$.

Example 7: The sequence s of Example 4 does not have a limit. For given any real number L, set $p = 1$. Then for M sufficiently large, $s_M = M > L + 1$; hence if $n > M$, s_n cannot be in $N(L, 1)$. Consequently, s cannot converge to L.

Example 8: We now prove that given any sequence s, $s_n \to 0$ if and only if $|s_n| \to 0$. We first note that for any $n \in \mathbb{N}$

$$|s_n - 0| = |s_n| = ||s_n|| = ||s_n| - 0|. \tag{6}$$

If $s_n \to 0$, then given any $p > 0$, there is M such that $n > M$ implies $|s_n - 0| < p$. From (6) we see that $n > M$ also implies $||s_n| - 0| < p$; hence $|s_n| \to 0$. On the other hand, if $|s_n| \to 0$, then given any $p > 0$ there is M such that $n > M$ implies $||s_n| - 0| < p$. Again from (6), we have $|s_n - 0| < p$ if $n > M$; hence $s_n \to 0$.

As an example of the use of the fact just proved, consider the sequence defined by $s_n = (-1)^n(1/n)$. Since $|s_n| = 1/n$ and it is known that $1/n \to 0$, we also have $(-1)^n(1/n) \to 0$.

We now give two other criteria for a sequence to converge.

Proposition 9: A sequence s converges to L if and only if given any $p > 0$ all but finitely many s_n are contained in $N(L, p)$.

We leave the proof of this proposition to the reader.

Proposition 10: A sequence s converges to L if and only if given any open set U which contains L all but finitely many s_n are elements of U.

PROOF: Suppose $s_n \to L$ and U is an open set which contains L. Then there is $p > 0$ such that $N(L, p) \subset U$ since U is open. On the other hand, since $s_n \to L$, all but finitely many s_n are in $N(L, p)$. Therefore all but finitely many s_n are in U.

Suppose that any open set which contains L contains all but finitely many s_n. Given any $p > 0$, $N(L, p)$ is an open set; hence $N(L, p)$ contains all but finitely many s_n. By Proposition 9, then $s_n \to L$.

The sequence of Example 3 converges to 0; one might wonder if it might converge to some other number as well. The next proposition tells us that any sequence can have at most one limit.

Proposition 11:   If s is a sequence which converges to both L and L', then $L = L'$.

PROOF: Suppose $L \neq L'$. Then by Proposition 8 there is $p > 0$ such that $N(L, p) \cap N(L', p) = \varnothing$. But since s converges to both L and L', $N(L, p)$ and $N(L', p)$ each contain all but finitely many s_n. This implies, however, that $N(L, p)$ and $N(L', p)$ must share at least one s_n in common, a contradiction. Therefore $L = L'$.

We note that the terms of the sequence of Example 3 not only approach 0 as n becomes large, but the terms of the sequence get closer to one another as well. We try to make this nebulous idea more precise in the following definition.

CAUCHY SEQUENCES

Definition 7:   *Let X, D be a metric space. A sequence s in X (that is, a function from $\mathbb{N}$ into X) is said to be a **Cauchy sequence** if given any $p > 0$, there is an integer M such that if m and n are any two integers greater than M, then $D(s_n, s_m) < p$.*

Less formally, a sequence s is a Cauchy sequence if given any $p > 0$, then any two terms of the sequence are less than distance p apart provided that the terms are chosen far enough along the sequence. A sequence s of real numbers is a Cauchy sequence if given $p > 0$ an integer M can be found such that if m and n are integers greater than M, then $|s_n - s_m| < p$.

Example 9:   The sequence of Example 3 is in fact a Cauchy sequence. For if $m > n$, then $|s_n - s_m| = |(m - n)/mn| < 1/n$. Since $1/n \to 0$, the sequence is a Cauchy sequence. It is no accident that this sequence is a Cauchy sequence as we see from the next proposition.

Proposition 12:   If s is a convergent sequence, then s is a Cauchy sequence.

PROOF: Suppose $s_n \to L$ and $p > 0$. Then there is an integer M such that $n > M$ implies $|s_n - L| < p/2$. If m and n are both greater than M, then $|s_n - s_m| \leq |s_n - L| + |s_m - L| < p/2 + p/2 = p$. Therefore s is a Cauchy sequence.

Although a convergent sequence must be a Cauchy sequence, it is not true for all metric spaces that a Cauchy sequence necessarily converges.

Example 10:   Let $\mathbb{Q}$ be the field of rational numbers with the absolute value metric, p-neighborhoods, open sets, and any related notions defined exactly as for $\mathbb{R}$. Pick a rational number s_n in $N(\sqrt{2}, 1/n)$ for each positive integer n (this is possible because every open interval contains a rational number).* It follows that $s_n \to \sqrt{2}$, which we have proved is not an element of $\mathbb{Q}$. Since $\{s_n\}$, $n \in \mathbb{N}$, converges in $\mathbb{R}$, it is a Cauchy sequence in $\mathbb{R}$, and, hence, is also a Cauchy sequence in $\mathbb{Q}$. Therefore $\{s_n\}$, $n \in \mathbb{N}$, is a Cauchy sequence in $\mathbb{Q}$ which does not converge in $\mathbb{Q}$.

Definition 8: *A metric space X with metric D is said to be* **complete** *if every Cauchy sequence in X converges to a point of X.*

A sequence $\{s_n\}$, $n \in \mathbb{N}$, in a metric space X, D is said to be **bounded** *if there is a positive number p and a point $x_0 \in X$ such that $D(s_n, x_0) < p$ for each $n \in \mathbb{N}$. (Informally, s is bounded if each s_n lies within distance p of some point of X.)*

COMPLETENESS OF $\mathbb{R}$

It is easy to show that a sequence s of real numbers is bounded if and only if there is a positive number K such that $s_n \in [-K, K]$ for each $n \in \mathbb{N}$; that is, all of the terms of s lie inside some closed interval containing 0.

Proposition 13: If s is a Cauchy sequence in $\mathbb{R}$, then s is bounded.

PROOF: Set $p = 1$. Since s is a Cauchy sequence, there is a positive integer M such that if m and n are integers greater than M, $|s_n - s_m| < 1$. Let K be the maximum of the numbers $|s_1|, |s_2|, \ldots, |s_M|$. Then for any positive integer n, $|s_n| < K + 1$; hence $-(K + 1) < s_n < K + 1$ for each $n \in \mathbb{N}$. Therefore s is bounded (each s_n is in $[-(K + 1), K + 1]$).

Since a sequence converges only if it is a Cauchy sequence, we also have the following corollary.

Corollary: Every convergent sequence in $\mathbb{R}$ is bounded.

The rational numbers with the absolute value metric do not form a complete metric space. The next proposition shows that $\mathbb{R}$, however, is complete.

Proposition 14: A sequence s of real numbers converges if and only if it is a Cauchy sequence.

---

* A Cauchy sequence of rational numbers which converges to $\sqrt{2}$ in $\mathbb{R}$ can also be defined explicitly by setting $s_1 = 1$ and $s_{n+1} = s_n + (1/8)(2 - s_n^2)$ (see Example 1 of Chapter 2).

PROOF: We have already proven that any convergent sequence is a Cauchy sequence. Suppose now that s is a Cauchy sequence. By Proposition 13 s is bounded; hence there is a positive number K such that $s_n \in [-K, K]$ for each $n \in \mathbb{N}$. Divide $[-K, K]$ into $[-K, 0]$ and $[0, K]$. Since there are infinitely many s_n, one of these intervals contains infinitely many s_n. Set the left end point of that interval equal to a_1 and the right end point of that interval equal to b_1.

Now divide $[a_1, b_1]$ into $[a_1, (1/2)(a_1 + b_1)]$ and $[(1/2)(a_1 + b_1), b_1]$. One of these intervals contains infinitely many s_n. Set the left end point of that interval equal to a_2 and the right end point equal to b_2. Continuing in like manner (that is, splitting the intervals which contain infinitely many s_n into two equal closed intervals, choosing a subinterval which contains infinitely many s_n, and labeling the end points appropriately), we obtain two sequences from the end points: the sequences $a_1, a_2, a_3, \ldots$ of left end points and $b_1, b_2, \ldots$ of right end points. By construction the following statements hold:

a) $a_{m-1} \leqslant a_m, m = 1, 2, 3, \ldots$
b) $b_m \leqslant b_{m-1}, m = 1, 2, 3, \ldots$
c) $b_m - a_m \leqslant (1/2)^m((K + 1) - (-(K + 1)) = (1/2)^{m-1}(K + 1)$,
   $m = 1, 2, 3, \ldots$

Let A be the set of a_m and B the set of b_m. Since A is non-empty and has an upper bound $K + 1$, A has a least upper bound, say a. Since B is non-empty and has a lower bound $-(K + 1)$, B has a greatest lower bound, say b. Then for each m, $a_m \leqslant a \leqslant b \leqslant b_m$. If $a \neq b$, then $b - a$ is positive. But

$$b - a \leqslant b_m - a_m = (1/2)^{m-1}(K + 1) \qquad (7)$$

for each m. If $b - a$ is positive, then there is a positive integer L such that if m is an integer larger than L, $(1/2)^{m-1}(K + 1) < b - a$, which is impossible by (7) and the fact that $(1/2)^{m-1} \to 0$ (Exercise 5). Therefore $a = b$.

We now show that our original sequence s converges to a. If $p > 0$, then there is an integer M_1 such that for any integer $m > M_1$, $b_m - a_m < p/2$. Consequently, $[a_m, b_m]$ is a subset of the $p/2$-neighborhood of a. Therefore $N(a, p/2)$ contains infinitely many of the s_n. Since s is a Cauchy sequence, there is a positive integer M_2 such that if n and m are integers greater than M_2, $|s_n - s_m| < p/2$. Since $N(a, p/2)$ contains infinitely many elements of s, there is at least one integer $m' > M_2$ such that $s_{m'} \in N(a, p/2)$, that is, $|s_{m'} - a| < p/2$. Therefore if $n > M_2$, then $|s_n - a| \leqslant |s_n - s_{m'}| + |s_{m'} - a| < p/2 + p/2 = p$. Thus $n > M_2$ implies $s_n \in N(a, p)$; hence $s_n \to a$.

The proof of Proposition 14 depended on the fact that every non-empty subset of $\mathbb{R}$ which has an upper bound has a least upper bound. We have seen that the order and metric properties of the real numbers are very closely related. The fact that *complete* is used to describe both the property that every Cauchy sequence converges and that any non-empty set which has an upper bound has a least upper bound suggests that the two types of completeness may be equivalent. Such is the case as we see in the next proposition. The next proposition also gives an important property of sequences of real numbers. We first introduce a definition.

Definition 9: *A sequence s of real numbers is said to be **monotone increasing** (**monotone decreasing**) if $s_n \leq s_{n+1}(s_n \geq s_{n+1})$ for each $n \in \mathbb{N}$.*

Proposition 15: The following statements are equivalent:

a) Every Cauchy sequence converges.
b) Every non-empty subset which has an upper bound has a least upper bound.
c) Any bounded monotone sequence converges.

PROOF: We have already seen in Proposition 14 that (b) implies (a). We now assume (a) and use it to prove (b). Let W be a non-empty subset of $\mathbb{R}$ such that W has an upper bound. We will find a least upper bound for W. We can find a sequence $s_1, s_2, s_3, \ldots$ such that $s_n \leq s_{n+1}$, each $s_n \in W$, and given any $w \in W$, there is a positive integer M such that $n > M$ implies $w \leq s_n$. There are a number of ways to construct such a sequence; for example, a system similar to that used to prove Proposition 14 might be used. The actual construction of such a sequence is left as an exercise. Such a sequence must be a Cauchy sequence. This can be proved as follows: Suppose s is not a Cauchy sequence. Then for some $p > 0$, there does not exist any integer M such that if n and m are greater than M, $|s_n - s_m| < p$. If $m < n$, then $s_m < s_n$. Therefore if $m < n$, $|s_m - s_n| = s_n - s_m$. Consequently, there is an integer m_1 such that $s_{m_1} - s_1 > p$. There is an integer $m_2 > m_1$ such that $s_{m_2} - s_{m_1} > p$; hence $s_{m_2} - s_1 > 2p$. There is $m_3 > m_2$ such that $s_{m_3} - s_{m_2} > p$; hence $s_{m_3} - s_1 > 3p$. Continuing in like fashion we see that s cannot have an upper bound. But since each $s_n \in W$, W cannot have an upper bound, a contradiction. Therefore s must be a Cauchy sequence.

Since s is a Cauchy sequence, s converges to some limit L. If L is not the least upper bound of W, then either L is not an upper bound of W, or L is an upper bound of W, but not the least upper bound. If L is not an upper bound of W, then there is $w \in W$ with $L < w$. Therefore there is a positive integer M for which $n > M$ implies $L < w \leq s_n$. Consequently, for $n > M$, $s_n \notin N(L, |w - L|)$, contradicting $s_n \to L$. Therefore L is an upper bound of W. If $L \neq \text{lub } W$, then there is an upper bound U of W with $U < L$. But then there is a positive integer M' such that $n > M'$ implies $s_n \notin N(L, |L - U|)$; hence again we contradict $s_n \to L$. Therefore $L = \text{lub } W$.

We next assume (b) and prove (c). Let s be a bounded monotone increasing sequence (the proof could just as well be carried out for a monotone decreasing sequence). Then $W = \{s_n \mid n \in \mathbb{N}\}$ is a non-empty set which is bounded above, and which therefore has a least upper bound L. We now show that $s_n \to L$. Choose $p > 0$. Since $L = \text{lub } W$, there is an integer M such that $s_M \in (L - p, L + p)$ (because there is s_M such that $L - p < s_M < L$). Since s is monotone increasing, if $n > M$, we have $L - p < s_M \leqslant s_n \leqslant L$; thus $s_n \in (L - p, L + p)$ for $n > M$. Consequently, $s_n \to L$.

Finally we prove that (c) implies (a). Let s be any Cauchy sequence. If s is monotone, then s converges. If s is not monotone, we can find a monotone sequence related to s in the following way: Let $t_1 = s_1$. Let t_2 be the first s_n after s_1 such that $s_1 \leqslant s_n$. Suppose t_n has been defined; let t_{n+1} be the first s_n after t_n (which is still a term of s) such that $t_n \leqslant t_{n+1}$. If it were not possible to find such a t_{n+1} for some n, this would imply s is monotone decreasing. The sequence $\{t_n\}$, $n \in \mathbb{N}$, has been defined inductively; each term of this sequence is also a term of s. However, $\{t_n\}$, $n \in \mathbb{N}$, is a monotone increasing sequence and hence has a limit L. We now show that $s_n \to L$.

Choose $p > 0$. Since s is a Cauchy sequence, there is an integer M such that whenever n and m are greater than M, $|s_n - s_m| < p/2$. Recall that each t_k is actually some s_n. Because of the manner in which the t_k are defined, there is an integer K such that t_K is an s_n for $n > M$. Moreover, since $t_k \to L$, we can find K so that $|t_K - L| < p/2$ as well. Therefore for $n > M$, we have

$$|s_n - L| \leqslant |s_n - t_K| + |t_K - L| < p/2 + p/2 = p.$$

It follows then that $s_n \to L$.

EXERCISES

1. Construct the sequence called for in the proof of Proposition 15.

2. Prove Proposition 9.

3. Prove that every irrational number is the limit of a sequence of rationals. Hint: Let a be an irrational number. For each positive integer n, choose a rational number s_n in $N(a, 1/n)$. Why is this choice possible? Prove $s_n \to a$.

4. a) Prove that any bounded monotone decreasing sequence converges.

b) Find a bounded monotone increasing sequence of rational numbers which does not converge to a rational number.

5. Determine whether or not the following sequences converge. If a sequence converges, determine its limit and prove the convergence. If a sequence does not converge, prove that it cannot converge to any real number.

a) $s_n = 1 - (1/n)$ d) $s_n = (-1)^n$

b) $s_n = 2/n^2$ e) $s_1 = 2, s_n = s_{n-1}/2$

c) $s_n = 1/n^{1/2}$ f) $s_1 = 0,$

$$s_n = (-1)^n(s_{n-1} + 1)/(n + 1)$$

6. Prove that if s is a sequence such that $s_n = b$, b a constant for $n > M$, M some integer, then $s_n \to b$.

7. Suppose s is a sequence which converges to L. Prove that $\{s_n \mid n \in \mathbb{N}\} \cup \{L\}$ is a closed set.

8. Prove that any Cauchy sequence in a metric space X, D is bounded. Prove that a sequence of real numbers is bounded if and only if each term of the sequence is contained in some closed interval containing 0.

3.3 CONTINUITY

We henceforth assume that all sets mentioned are subsets of $\mathbb{R}$ unless a contrary assumption is stated explicitly.

We now turn our attention to functions from $\mathbb{R}$ into $\mathbb{R}$—but not just any functions. If $f: W \to \mathbb{R}$ (where $W \subseteq \mathbb{R}$) then to justify its discussion in this text, f must somehow be related to the structure of $\mathbb{R}$. Since $\mathbb{R}$ has a great deal of structure—algebraic, order, and metric—there are many properties that f might have that relate it to one or more of the structures of $\mathbb{R}$. For example, f might be order-preserving, or distance-preserving, or it might respect the algebraic operations of $\mathbb{R}$. In this section we concern ourselves with those functions which "respect" the metric structure of $\mathbb{R}$. To say that a function $f: \mathbb{R} \to \mathbb{R}$ must preserve distances, that is, $|r - r'| = |f(r) - f(r')|$, in order to be a metric-structure-respecting function is actually to demand too much of f. For example, the function f defined by

$f(x) = 2x$ looks like it should be "respectable," from almost any point of view, but it does not preserve distances. Informally, what we will ask f to do is to map numbers which are "close together" into numbers which are "close together." The function $f(x) = 2x$ seems to have this property since it takes any two numbers which are distance p apart into numbers distance $2p$ apart. If p is small, then $2p$ will also be small. We make this rather intuitive notion of "keeping numbers close together" more precise in the following definition.

CONTINUITY

Definition 10:   *A function f from a subset W of* $\mathbb{R}$ *into* $\mathbb{R}$ *is said to be* **continuous at a point** *a of W if given any* $p > 0$, *there is* $q > 0$ *such that whenever* $x \in W$ *and* $|a - x| < q$, *then* $|f(a) - f(y)| < p$. *If f is not continuous at a, then we say that f is* **discontinuous** *at a.*

We say that f is continuous on W if f is continuous at each point of W.

It is easy to verify that according to Definition 10 a sequence s (recall that $s: \mathbb{N} \to \mathbb{R}$) is continuous.

In elementary calculus the discussion of continuity at a is usually restricted to the situation in which some open interval containing a is a subset of the domain of the function being considered.

Informally, Definition 10 tells us that $f: W \to \mathbb{R}$ is continuous at $a \in W$ if given any $p > 0$, in order to have $f(x)$ p-close to $f(a)$ it suffices to have x q-close to a for a suitable $q > 0$; in other words, f takes points which are close to a into points which are close to $f(a)$. The following examples illustrate Definition 10.

Example 11:   Consider the function $f: \mathbb{R} \to \mathbb{R}$ defined by $f(x) = 3x + 1$ for each $x \in \mathbb{R}$. Then f is continuous at 3. For let $p > 0$ and set $q = p/3$. Then if $|3 - x| < q = p/3$, we have $3 |3 - x| = |3^2 - 3x| = |(3 \cdot 3 + 1) - (3x + 1)| = |f(3) - f(x)| < p$. Hence f satisfies Definition 10 at 3. Less formally, given any $p > 0$, $3x + 1$ is p-close to $f(3)$ provided that x is $p/3$-close to 3. Actually, f is continuous on all of $\mathbb{R}$; for given any $p > 0$ and $a \in \mathbb{R}$, setting $q = p/3$, we find that $|a - x| < q$ implies $|f(a) - f(x)| < p$.

Example 12:   Consider the function $f: \mathbb{R} \to \mathbb{R}$ defined by $f(x) = x^2$. We now prove that f is continuous for any $a > 0$. (It can be shown that f is continuous for any $a \in \mathbb{R}$.) For $a > 0, f(a) = a^2$ (refer to Fig. 4). Choose any $p > 0$; we can suppose that $p < a^2$ so that $a^2 - p$ is positive. Set $q = \text{minimum} \ (a - \sqrt{a^2 - p}, \ \sqrt{a^2 + p} - a)$. If $|a - x| < q$, then

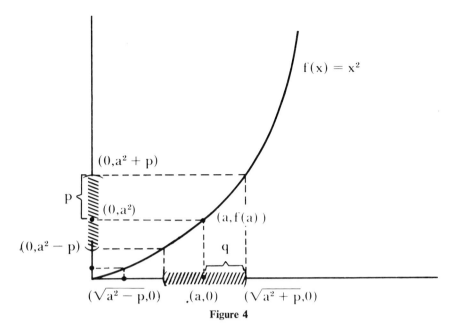

$(0, a^2 + p)$

p

$(0, a^2)$

$(a, f(a))$

$f(x) = x^2$

$(0, a^2 - p)$

q

$(\sqrt{a^2 - p}, 0)$ $(a, 0)$ $(\sqrt{a^2 + p}, 0)$

Figure 4

$\sqrt{a^2 - p} < x < \sqrt{a^2 + p}$. Consequently, $a^2 - p < x^2 = f(x) < a^2 + p$, or $|f(a) - f(x)| < p$. Therefore f is continuous at a.*

Example 13: Consider the function $f: \mathbb{R} \to \mathbb{R}$ defined by

$$f(x) = \begin{cases} 0, & x < 1, \\ 1, & x \geqslant 1. \end{cases}$$

Figure 5 is the graph of f. Although f is continuous at every real number except 1; f is discontinuous at 1. For set $p = 1/2$. Since there are real

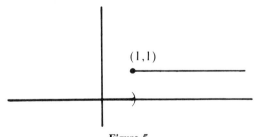

$(1,1)$

Figure 5

---

* The proof might also have proceeded as follows: For $a > 0$, $f(a) = a^2$. Choose any $p > 0$ and set $q = \min\{1, p/(2a + 1)\}$. Now $|x + a| < 1 + 2a$ (why?). Therefore if $|x - a| < q$, we have $|f(x) - a^2| = |x^2 - a^2| = |x - a| |x + a| < q(1 + 2a) \leqslant p$. Therefore f is continuous at a.

numbers less than 1 which are arbitrarily close to 1, we cannot find any $q > 0$ such that $|x - 1| < q$ implies $|f(1) - f(x)| = |1 - f(x)| < 1/2$. For any $q > 0$, let x be less than 1 and within distance q of 1 (e.g., let $x = 1 - q/2$). Then $|1 - f(x)| = |1| = 1 \nless 1/2$. Therefore f is discontinuous at 1.

Restating Definition 10 in terms of neighborhoods, we have the following proposition.

Proposition 16:   A function $f: W \to \mathbb{R}$ is continuous at $a \in W$ if and only if given any $p > 0$, there is $q > 0$ such that $f(N(a, q) \cap W) \subseteq N(f(a), p)$.

Corollary: A function $f: W \to \mathbb{R}$ is continuous on W if and only if given any $a \in W$ and any $p > 0$, there is $q > 0$ (which generally depends on a) such that $f(N(a, q) \cap W) \subseteq N(f(a), p)$.

CONTINUITY AND OPEN SETS

Since open sets are characterized in terms of neighborhoods, we can also characterize continuity in terms of open sets.

Proposition 17:   A function $f: W \to \mathbb{R}$ is continuous if and only if given any open set U of $\mathbb{R}$, $f^{-1}(U)$ is the intersection of W with an open subset of $\mathbb{R}$.

PROOF: Suppose that given any open set U of $\mathbb{R}$, $f^{-1}(U) = V \cap W$, where V is open. Let $a \in W$ and $p > 0$. Now $N(f(a), p)$ is an open set; hence $f^{-1}(N(f(a), p)) = V \cap W$, where V is open. Since V is open, there is $q > 0$ such that $N(a, q) \subseteq V$. But then $N(a, q) \cap W \subseteq f^{-1}(N(f(a), p))$. Therefore $f(N(a, q) \cap W) \subseteq N(f(a), p)$. Therefore f is continuous by Proposition 16.

Suppose that $f: W \to \mathbb{R}$ is continuous and U is an open set. Consider $a \in f^{-1}(U) \cap W$ (it is possible for this intersection to be empty, but then it is also open). Since U is open, there is $p > 0$ such that $N(f(a), p) \subseteq U$. Since f is continuous, there is $q_a > 0$ such that $f(N(a, q_a) \cap W) \subseteq N(f(a), p) \subseteq U$. Therefore $N(a, q_a) \cap W \subseteq f^{-1}(U)$. Set $V = \bigcup \{N(a, q_a) \mid a \in W\}$. Then V is an open set. Moreover, $f^{-1}(U) = W \cap V$. This completes the proof of the proposition.

A criterion for continuity similar to, but sometimes more practically workable than, that given in Proposition 17 is given in the following proposition.

Proposition 18:   The function $f: W \to \mathbb{R}$ is continuous if and only if given any open interval $(a, b) \subseteq \mathbb{R}$, $f^{-1}((a, b))$ is the intersection of W with an open set.

PROOF: If f is continuous, then since any open interval is an open set, $f^{-1}((a, b))$ is the intersection of W with an open set by the previous proposition.

Suppose that given any open interval (a, b), $f^{-1}((a, b)) = W \cap V$, where V is open, and let U be any open set. Then U is the union of open intervals (Proposition 6), say $U = \bigcup_I H_i$, where each H_i is an open interval. Then $f^{-1}(U) = \bigcup_I f^{-1}(H_i) = \bigcup_I (W \cap V_i) = W \cap (\bigcup_I V_i)$, where $f^{-1}(H_i) = W \cap V_i$, V_i some open set. Since $\bigcup_I V_i$ is open because it is the union of a family of open sets, $f^{-1}(U)$ is the intersection of W with an open set. Therefore f is continuous by the previous proposition.

Since we have criteria for both continuity of a function and convergence of a sequence in terms of open sets, we might feel there ought to be a criterion for continuity in terms of convergence of sequences. The next proposition gives such a criterion.

CONTINUITY AND SEQUENCES

Proposition 19:   The function $f: W \to \mathbb{R}$ is continuous at $a \in W$ if and only if given any sequence $\{s_n\}$, $n \in \mathbb{N}$, in W, such that $s_n \to a$, $f(s_n) \to f(a)$.

PROOF: Suppose that f is continuous and $s_n \to a$. We must show $f(s_n) \to f(a)$. Let $p > 0$. Since f is continuous at a there is $q > 0$ such that $f(N(a, q) \cap W) \subseteq N(f(a), p)$ (Proposition 16). Since $s_n \to a$ and each $s_n \in W$, all but finitely many s_n are in $N(a, q) \cap W$. Therefore all but finitely many $f(s_n)$ are in $f(N(a, q)) \subseteq N(f(a), p)$. Consequently, $f(s_n) \to f(a)$.

Suppose that $f(s_n) \to f(a)$ whenever $s_n \to a$, but f is not continuous at a. Since f is not continuous at a, there is $p > 0$ such that $f(N(a, q) \cap W) \not\subseteq N(f(a), p)$ for any $q > 0$. Set $q_n = 1/n$, $n = 1, 2, 3, \ldots$ For each n select s_n from $N(a, q_n) \cap W$ such that $f(s_n) \notin N(f(a), p)$. Then $s_n \to a$ (the proof of this fact is left as an exercise), but $f(s_n)$ does not converge to $f(a)$ since no $f(s_n)$ is contained in $N(f(a), p)$, and this contradicts our hypothesis about f. Therefore f must be continuous, and the proposition is proved.

We now illustrate the use of Propositions 16 to 19 by means of several examples.

Example 14:   The function $f: \mathbb{R} \to \mathbb{R}$ defined by $f(x) = 2x$ is continuous. For if (a, b) is any open interval in $\mathbb{R}$, then $f^{-1}((a, b)) = (a/2, b/2)$ which is also an open interval, and hence an open set. Therefore f is continuous by Proposition 18.

Example 15:   Let $f: \mathbb{R} \to \mathbb{R}$ be defined by $f(x) = x^2$, and (a, b) be an open interval. If a and b are both non-positive, then $f^{-1}((a, b)) = \emptyset$ is the intersection of $\mathbb{R}$ with the open set $\emptyset$. If a is non-positive and b is positive, then $f^{-1}((a, b)) = (-\sqrt{b}, \sqrt{b})$, an open interval. If a and b are both positive, then $f^{-1}((a, b)) = (-\sqrt{b}, -\sqrt{a}) \cup (\sqrt{a}, \sqrt{b})$, again the union of open intervals, and, hence, open. Therefore f is continuous by Proposition 18.

Example 16:   Let $f: \mathbb{R} \to \mathbb{R}$ be defined by $f(x) = 2$ for $x \neq 0$, but $f(0) = 0$. The graph of f is given in Figure 6. The sequence $s_n = 1/n$

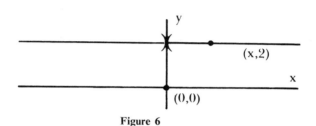

Figure 6

converges to 0, but the sequence $f(s_n) = 2$ converges to 2 rather than $f(0) = 0$. Consequently, this function f is not continuous at 0 by Proposition 19. If we redefined $f(0)$ to be 2 instead of 0, then the redefined function would be continuous on all of $\mathbb{R}$.

Despite the propositions we have proved, it is often a long and arduous job to prove that certain functions are continuous using only what we have so far. For example, the function f defined $f(x) = x^{34} - 35x^{27} + 14x^{19} + 1$ is continuous, but a formal proof at the moment would be rather lengthy. Fortunately, we will have a number of other tools for investigating continuity later in this text. We conclude this section with a proposition which gives but one such tool.

Proposition 20:   If $f: W \to \mathbb{R}$ and $g: W' \to \mathbb{R}$ are both continuous, then $f \circ g$ is continuous on the set on which it is defined.

PROOF: Suppose $\{s_n\}$, $n \in \mathbb{N}$, is a sequence in the domain of $f \circ g$ such that $s_n \to a$ for any a for which $(f \circ g)(a)$ is defined. Then since g is continuous, $g(s_n) \to g(a)$. Therefore since f is continuous $f(g(s_n)) = (f \circ g)(s_n) \to f(g(a)) = (f \circ g)(a)$. This establishes the continuity of $f \circ g$ by Proposition 19.

Example 17:   Since $f(x) = x^2$ is continuous, $(f \circ f)(x) = (x^2)^2 = x^4$ is also continuous.

EXERCISES

1. Prove Proposition 16.

2. Prove that the sequence $\{s_n\}$, $n \in \mathbb{N}$, constructed in the second half of the proof of Proposition 19 does in fact converge to a.

3. Each of the following functions is intended to be defined for as much of $\mathbb{R}$ as makes sense for the particular function. Determine which of these functions are continuous. If a function fails to be continuous, indicate at which points the function is discontinuous; try to redefine the function at the points of discontinuity so as to have a continuous function.

a) $f(x) = x^3$

b) $f(x) = 3x + 1$

c) $f(x) = x^{1/2}$

d) $f(x) = \begin{cases} |x|, & \text{for } x < 0, \\ x^2, & \text{for } x \geqslant 0 \end{cases}$

e) $f(x) = \begin{cases} x, & \text{for } x \leqslant 0, \\ 1, & \text{for } 0 < x < 2, \\ x + 1, & \text{for } 2 \leqslant x \end{cases}$

4. Formulate a proof of Proposition 20 using Proposition 16.

5. Prove that $f: W \to \mathbb{R}$ is continuous if and only if given any closed set F, $f^{-1}(F)$ is the intersection of W and a closed set. Prove or disprove: $f: W \to \mathbb{R}$ is continuous if and only if given any closed interval $[a, b]$, $f^{-1}([a, b])$ is the intersection of W and a closed set.

6. A set W is said to be *bounded* if it is a subset of a closed interval. Prove that if $f: \mathbb{R} \to \mathbb{R}$ is continuous and W is bounded, then $f(W)$ is also bounded.

7. Prove that any sequence of real numbers is a continuous function.

3.4 COMPACTNESS

One of the most important notions in mathematics is *compactness*. It is not the purpose of this text to go into a history of the evolution of this fundamental concept; nevertheless, compactness plays an important role not only in real analysis, but in virtually all areas of mathematical study.

We will define compactness in terms of *open coverings* (even though this was not the way it was defined until fairly recently in mathematical history) because this clearly points up the topological nature of compactness, enables the reader to generalize more easily much of what we will do only in the rather limited context of the real numbers, and is also one of the most elegant and workable characterizations of the property that compactness is intended to convey.

Definition 11:   *A collection $\{U_i\}$, $i \in I$, of open sets is said to be an **open cover** (or **open covering**) of W if each element of W is contained in at least one U_i.   The open sets $\{V_j\}$, $j \in J$, are said to be a **subcover** (or **subcovering**) relative to $\{U_i\}$, $i \in I$, or simply a **subcover** if no ambiguity will result, if each V_j is also a U_i and $\{V_j\}$, $j \in J$, is also an open cover of W.*

*A subset A of $\mathbb{R}$ is said to be **compact** if every open cover of A has a finite subcover of A, that is, given any open cover of A, finitely many members of the cover are all that are needed to cover A.*

Any subset A of $\mathbb{R}$ generally has many open coverings, for example, $\{N(x, p)\}$, $x \in A$, is a covering for any $p > 0$. To be compact, A has to be such that *every* open covering of A has a finite subcovering.

Example 18:   The family $\{(-n, n)\}$, $n \in \mathbb{N}$, of open intervals is an open cover of $\mathbb{R}$. Since no finite number of members of this covering of $\mathbb{R}$ can cover $\mathbb{R}$, we conclude that $\mathbb{R}$ is not compact.

Example 19:   Let s be a sequence and suppose $s_n \to L$. Then $A = \{s_n \mid n \in \mathbb{N}\} \cup \{L\}$ is compact. For suppose $\{U_i\}$, $i \in I$, is an open cover of A. Because the U_i form a cover of A, each point of A is in at least one (and possible more than one) of the U_i. In particular, $L \in U_{i'}$ for some $i' \in I$. Since $s_n \to L$, $U_{i'}$ contains all but finitely many of the terms of s. Consequently, $U_{i'}$ together with at most finitely many more members of the cover are required to cover all of A. Since $\{U_i\}$, $i \in I$, contains a finite subcover of A, A is compact.

Although Definition 11 is one of the most elegant characterizations of compactness, since it is also one of the most abstract, it does not give a clear indication why the notion of compactness is important. We therefore spend a moment "motivating" this central concept.

The simplest kind of subset of any set is a finite subset. A compact subset of $\mathbb{R}$ is "almost finite" in the following sense: Given any collection of arbitrarily small open intervals which cover $\mathbb{R}$, it takes only finitely many of the intervals to cover any compact subset of $\mathbb{R}$.* More specifically, many

---

* A prominent contemporary American mathematician E. Hewitt has the following footnote in "The Role of Compactness in Analysis," *American Mathematical Monthly*, Vol. 67, No. 6 (May, 1960): The following paraphrase of [the definition of *compact* in Definition 11] is attributed to Hermann Weyl (1885–1955). "If a city is compact, it can be guarded by a finite number of arbitrarily near-sighted policemen."

theorems about finite sets or theorems stated in finite settings are true as well under a compactness instead of a finiteness hypothesis, and sometimes the proof of the more general result can be effected with only slight modifications of the original proof.

Although there are a great many compact subsets of $\mathbb{R}$, the ones we will consider most often are closed intervals. We now prove that the closed interval $[0, 1]$ is compact.

Proposition 21: $[0, 1]$ is compact.

PROOF: Let $\{U_i\}$, $i \in I$, be any open cover of $[0, 1]$. Let

$$T = \{x \in [0, 1] \mid \text{finitely many of the } U_i \text{ cover } [0, x)\}.$$

Then $T \neq \varnothing$, and 1 is an upper bound for T. Therefore T has a least upper bound, say u. If $u = 1$, then we are done (since finitely many of the U_i cover $[0, 1)$, and hence at most one more of the U_i will be needed to obtain a finite cover of $[0, 1)$). Suppose then that $0 \leqslant u < 1$. Then either $u \in T$, or $u \notin T$.

Case 1: $u \in T$. Then finitely many of the U_i, say $U_{i_1}, \ldots, U_{i_n}$ cover $[0, u)$. There is, however, $U_{i'}$ such that $u \in U_{i'}$; therefore $U_{i'}, U_{i_1}, \ldots, U_{i_n}$ is an open cover of $[0, u]$. It is then clear (Fig. 7) that u could not be an

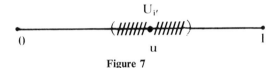

Figure 7

upper bound for T.

Case 2: $u \notin T$. Then there is U_j such that $u \in U_j$ and finitely many of the U_i do not cover $[0, u) \sim U_j$. Therefore u is not the least upper bound for T.

Since both cases lead to contradictions, we could not have $0 \leqslant u < 1$. Therefore $u = 1$; hence $[0, 1]$ is compact.

An argument entirely similar to that used to prove Proposition 21 can be used to prove Proposition 22.

Proposition 22: Any closed interval is compact.

Proposition 23: Any compact set is closed.

PROOF: Suppose W is compact and $y \notin W$. We will show that y is contained in an open set V_y which does not contain a point of W. From this it follows that $\mathbb{R} \sim W$ is the union of all such V_y for all $y \in \mathbb{R} \sim W$, and thus being the union of a family of open sets, it is open. This in turn implies that W is closed. For each $w \in W$, there are disjoint open sets H_w and G_w such that $w \in H_w$ and $y \in G_w$. (cf. Proposition 8). Then $\{H_w\}$, $w \in W$, is an open cover of W. Since W is compact, finitely many H_w, say $H_{w_1}, \ldots, H_{w_n}$, cover W. Set $U = H_{w_1} \cup \ldots \cup H_{w_n}$ and $V_y = G_{w_1} \cap \ldots \cap G_{w_n}$. Then $W \subseteq U$, $y \in V_y$, and $U \cap V_y = \varnothing$. Therefore $W \cap V_y = \varnothing$, and the proposition is proved.

Proposition 24:   Any closed subset of a compact set is compact.

PROOF: Let A be a closed subset of a compact set W and suppose $\{U_i\}$, $i \in I$, is an open cover of A. Then $\mathbb{R} \sim A$ together with the U_i's forms an open cover of W. This cover in turn contains a finite subcover of W. This implies, however, that finitely many of the U_i cover A; therefore A is compact.

The notion of a bounded subset of $\mathbb{R}$ was presented in Definition 8 of Chapter 2. The next proposition gives another characterization of boundedness.

Proposition 25:   A set W is bounded if and only if there is a positive number K such that $W \subseteq [-K, K]$.

We leave the proof of Proposition 25 to the reader.

Proposition 26:   Any compact set W is bounded.

PROOF: The open cover $\{(-n, n)\}$, $n \in \mathbb{N}$, of W, contains a finite subcover of W. Therefore $W \subseteq (-n, n)$ for some $n \in \mathbb{N}$. But then $W \subseteq [-n - 1, n + 1]$; hence W is bounded.

We have thus far shown that any compact set is both closed and bounded. We now prove that being closed and bounded suffices to make a set compact.

Proposition 27 (*The Heine-Borel Theorem*):   Any closed bounded set is compact.

PROOF: If W is closed and bounded, then W is a closed subset of a compact set $[-K, K]$ for some $K > 0$. Therefore W is compact by Proposition 24.

Another important property of compact sets is expressed in terms of *limit points*.

Definition 12: *A· point x is said to be a **limit point**, or **accumulation** point, of a set A if any open set which contains x contains at least one point of A other than x.*

The notion of limit point of a set is not to be confused with the limit of a sequence.

Example 20: Consider the sequence defined by $s_n = 1$. Then $s_n \to 1$, but 1 is not a limit point of $A = \{s_n \mid n \in \mathbb{N}\}$ since no open set which contains 1 contains a point of A other than 1 (1 is the only point of A).

Example 21: Suppose w is a limit point of A. We now show that there is a sequence of points of A which converges to w. For each $n \in \mathbb{N}$, $N(w, 1/n)(= (w + 1/n, w - 1/n))$ contains a point s_n of A other than w. We leave it to the reader to prove that $s_n \to w$.

Proposition 28: If W is a compact set, then any infinite subset A of W has a limit point which is an element of W.

PROOF: Suppose A does not have a limit point in W. Then for each $x \in W$ there is an open set U_x which contains at most one point of A. Therefore $\{U_x\}$, $x \in W$, must be an open cover of W without a finite subcover since any finite subcover can contain only finitely many points of A.

Example 22: Since the integers form an infinite subset of $\mathbb{R}$ which does not have a limit point, we confirm again that $\mathbb{R}$ is not compact.

If every infinite subset of a set W has a limit point in W, then W is compact; that is, the converse of Proposition 28 is true, but we will not prove this until later.

EXERCISES

1. A family $\{F_i\}$, $i \in I$, of sets is said to have the *finite intersection property* (FIP) if any finite number of the F_i have a non-empty intersection. Prove that a set W is compact if and only if every family of closed subsets of W having the FIP has a non-empty intersection. Show directly that $\mathbb{R}$ does not have the FIP.

2. Prove that a set F is closed if and only if every sequence of points of F which converges, converges to a point of F.

3. Prove a set F is closed if and only if it contains all of its limit points.

4. Prove that any closed interval is compact. Prove that any finite union of compact sets is compact. Prove that the intersection of any family of compact sets is compact.

5. Let $\{s_n\}$, $n \in \mathbb{N}$, be any sequence such that $s_n \to L$. Under what conditions is L a limit point of $\{s_n \mid n \in \mathbb{N}\}$?

6. Determine which of the following sets are compact. Prove all of your assertions.

a) $(0, 1)$ d) $\{x \mid x^{24} + x^{18} - 8x^6 - 19 = 0\}$
b) $[0, \infty)$ e) $\{x \mid x^3 > 0\} \cap \{x \mid x^2 < 1\}$
c) $[0, 1] \cup \{3\}$

7. Prove that any non-empty compact set contains its lub and glb.

3.5   COMPACTNESS AND CONTINUOUS FUNCTIONS. CONNECTEDNESS

COMPACTNESS AND FUNCTIONS

We have already briefly examined continuous functions from a subset W of $\mathbb{R}$ into $\mathbb{R}$. If the set W is compact, then continuous functions $f: W \to \mathbb{R}$ enjoy a number of special properties.

Proposition 29:   If $f: W \to \mathbb{R}$ is continuous and W is compact, then $f(W)$ is compact.

PROOF: Suppose $\{U_i\}$, $i \in I$, is an open cover of $f(W)$. Since f is continuous, $f^{-1}(U_i) = W \cap V_i$, for each $i \in I$, where V_i is open. Therefore $\{V_i\}$, $i \in I$, forms an open cover of W. But W is compact, hence finitely many V_i, say $V_{i_1}, V_{i_2}, \ldots, V_{i_n}$ cover W. Therefore $U_{i_1}, \ldots, U_{i_n}$ cover $f(W)$; hence $f(W)$ is compact.

Corollary: If $f: W \to \mathbb{R}$ is continuous and W is compact, then $f(W) = \{f(x) \mid x \in W\}$ is bounded and hence has a lub M and glb m. Moreover, there are a and b in W such that $f(a) = m$ and $f(b) = M$. (That is, f actually assumes a least value and a greatest value on W.)

PROOF: The corollary follows directly from Proposition 29 and Exercise 7 of the preceding section.

Definition 13:   *A function $f: W \to \mathbb{R}$ is said to be **bounded** if $f(W)$ is a bounded set.*

Thus, the corollary to Proposition 29 tells us, among other things, that if W is compact and $f: W \to \mathbb{R}$ is continuous, then f is bounded. It is possible for a function to be bounded without being continuous. If $f: W \to \mathbb{R}$ is not continuous, then f may not be bounded even if W is compact. We illustrate these points in the following examples.

Example 23:   The function $f: (0, 1) \to \mathbb{R}$ defined by $f(x) = x$ is bounded even though $(0, 1)$ is not compact. Note, however, that f assumes neither a maximum or minimum value on $(0, 1)$.

Example 24:   Define $g: [-1, 1] \to \mathbb{R}$ by $g(x) = 1/x$ if $x \neq 0$ and $g(0) = 0$. The graph of g is given in Figure 8. This function is not bounded,

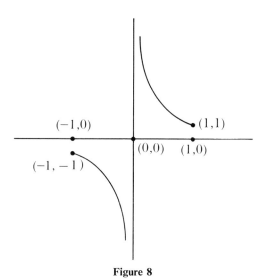

Figure 8

even though $[-1, 1]$ is compact.

In the corollary to Proposition 29 it is not necessary that a and b be either lub W or glb W. Consider the following example.

Example 25:   Let $f: [-1/2, 3/2] \to \mathbb{R}$ be defined by

$$f(x) = \begin{cases} |x|, & \text{for } x \leqslant 1 \\ 2 - |x|, & \text{for } 1 < x. \end{cases}$$

The graph of f is given in Figure 9. Note that the maximum and minimum values of f do not occur at the end points of $[-1/2, 3/2]$.

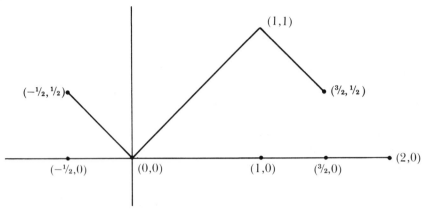

Figure 9

UNIFORM CONTINUITY

The next proposition is used in the proof of Proposition 31.

Proposition 30:   Let W be compact and $\{U_i\}$, $i \in I$, be an open cover of W. Then there is $p > 0$ such that for any $x \in X$ $N(x, p) \subseteq U_i$ for some i. That is, there is $p > 0$ such that the p-neighborhood of every point of W is a subset of some member of the open cover. Such a number p is called a *Lebesgue number* of the cover and is dependent on the cover for its value.

PROOF: Each element x of W is in at least one U_i because the U_i's cover W. Since each U_i is open, for each $x \in W$ we may select p_x such that $N(x, p_x)$ is a subset of at least one of the U_i. Since a selection has been made for each x, $\{N(x, p_x/2)\}$, $x \in W$, is an open cover of W. Since W is compact, we can find a finite number of elements of W, say $x_1, \ldots, x_n$, such that $N(x_1, p_{x_1}/2), \ldots, N(x_n, p_{x_n}/2)$ form an open cover of W. Let p be the minimum of $p_{x_1}/2, \ldots, p_{x_n}/2$. We now show that p is a Lebesgue number for the original open cover of W.

If $x \in W$, then $x \in N(x_j, p_{x_j}/2)$ for some $j = 1, \ldots, n$. If $y \in N(x, p)$, then $|y - x_j| \leqslant |y - x| + |x - x_j| < p + p_{x_j}/2 \leqslant p_{x_j}$. Therefore $N(x, p) \subseteq N(x_j, p_{x_j}) \subseteq U_i$ for some i.

We recall that the definition of continuity of $f: W \to \mathbb{R}$ can be expressed: Given any $w \in W$ and $p > 0$, there is $q > 0$ such that $f(N(w, q) \cap W) \subseteq N(f(w), p)$. Note that while p can be chosen arbitrarily, q is dependent on *both* w and p. Even if the same p is used for each $w \in W$, it may not be possible to find a $q > 0$ which will work for every $w \in W$. This point is illustrated in the following example.

Example 26: Let $W = \{x \mid 0 < x\}$ and $f(x) = 1/x$ for each $x \in W$. It is easily verified that f is continuous. It is fairly clear (and may be proven rigorously rather easily) that for any $p > 0$ and any $q > 0$ there is $x \in W$ such that $f(N(x, q))$ is not a subset of $N(f(x), p)$ (Fig. 10).

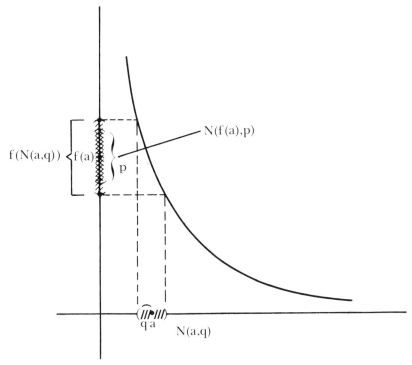

Figure 10

Note that W in Example 26 is not compact. If W were compact, then we could find a q for each p which was independent of x, as we will see from the next proposition.

Definition 14: *Let $f: W \to \mathbb{R}$. Then f is said to be **uniformly continuous** if given any $p > 0$, there is $q > 0$ such that for **any** $w \in W, f(N(x, q) \cap W) \subseteq N(f(x), p)$. (That is, q can be found independent of x.)*

Proposition 31: If $f: W \to \mathbb{R}$ is continuous and W is compact, then f is uniformly continuous.

PROOF: Choose any $p > 0$. Then $\{N(f(w), p/2)\}$, $w \in W$, is an open cover of $f(W)$. Since f is continuous, $f^{-1}(N(f(w), p/2)) = W \cap V_{f(w)}$ for each $f(w) \in f(W)$, where $V_{f(w)}$ is open. Then $\{V_{f(w)}\}$, $f(w) \in f(W)$, is an open cover of W. Let q be a Lebesgue number of this open cover. We leave it to the reader to prove that $f(N(w, q)) \subseteq N(f(w), p)$ for each $w \in W$. Therefore f is uniformly continuous.

Corollary: Any function from a closed interval of $\mathbb{R}$ into $\mathbb{R}$ is uniformly continuous provided that it is continuous.

The set W need not be compact in order for a continuous function $f: W \to \mathbb{R}$ to be uniformly continuous, as we see from the next example.

Example 27:   Let $W = (0, 1)$ and let $f: W \to \mathbb{R}$ be defined by $f(x) = 7x + 1$. Then f is uniformly continuous. For any $p > 0$, $q = p/7$ insures that $f(N(x, q) \cap W) \subseteq N(f(x), p)$; hence f is uniformly continuous even though W is not compact.

CONNECTEDNESS

Another important topological property (that is, a property related to the open sets) to be considered in addition to compactness is *connectedness*.

Definition 15:   *A subset W of $\mathbb{R}$ is said to be **disconnected** if there are open sets U and V such that $W \subset U \cup V$, $W \cap U$ and $W \cap V$ are both non-empty, and $(U \cap V) = \varnothing$ (Fig. 11). If W is not disconnected, we say that W is **connected**.*

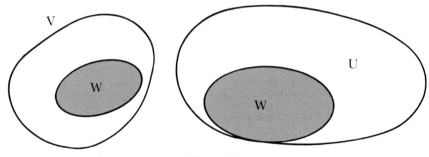

Figure 11

Figure 11 gives some indication of why the term "disconnected" is applied to W.

Proposition 32:   The closed interval $[0, 1]$ is connected.

PROOF: Suppose $[0, 1]$ is disconnected. Then there are open sets U and V such that $[0, 1] \subseteq U \cup V$ and U and V contain no points of $[0, 1]$ in common. Suppose u and v are points of $[0, 1]$ with $u \in U$ and $v \in V$. We may assume $u < v$ (relabeling U and V, if necessary). Let S be the set of numbers s such that $s \leqslant u$ or $[u, s] \subseteq U$. Then S has 1 as an upper bound, and hence S has a least upper bound a with $0 < a < 1$. Note that S has been defined so that for any t with $u < t < a$, we have $[u, t] \subseteq U$. Since any element of $[0, 1]$ is either in U or V, either $a \in U$ or $a \in V$. Suppose $a \in U$. Since U is open, there is $p > 0$ such that $(a - p, \; a + p) \subseteq U$. Then $[a - p/2, a + p/2] \subseteq U$ (Fig. 12); hence $a + p/2 \in S$. This, however, contradicts the assumption that a is an upper bound for S. Suppose then that

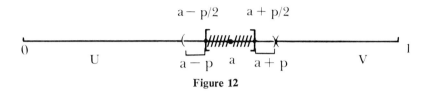

Figure 12

$a \in V$. Then there is $p > 0$ such that $N(a, p) \subseteq V$, and thus $a - p/2$ is an upper bound of S contradicting $a = \text{lub } S$. Therefore $[0, 1]$ is not disconnected; hence it must be connected.

An entirely similar argument can be used to prove the next proposition.

Proposition 33: Closed intervals, open intervals, half-open intervals, open half-lines, closed half-lines, and $\mathbb{R}$ itself are connected.

Proposition 34: If W is a connected subset of $\mathbb{R}$ and a and b are elements of W with $a < b$, then $[a, b] \subseteq W$.

PROOF: Suppose $r \in [a, b]$, but $r \notin W$. Since a and b are in W, $a < r < b$. Let U and V be the two open half-lines with r as their end point. Then U and V separate W into disjoint pieces. It follows that W is disconnected, contradicting the assumption that W is connected. Therefore $[a, b]$ must be a subset of W.

Corollary 1: The only connected subsets of $\mathbb{R}$ in addition to those mentioned in Proposition 33 are $\varnothing$ and one point subsets.

The proof of this corollary is left to the reader.

Corollary 2: The only compact connected subsets of $\mathbb{R}$ are closed intervals, $\varnothing$, and one point subsets.

Proposition 35: If $f: W \to \mathbb{R}$ is continuous and W is connected, then $f(W)$ is also connected.

PROOF: Suppose $f(W)$ is not connected. Then there are open sets U and V whose union contains $f(W)$, but which share no points of $f(W)$ in common. Since f is continuous $f^{-1}(U) = W \cap U'$ and $f^{-1}(V) = W \cap V'$, where U' and V' are also open; moreover, W is contained in $U' \cup V'$ and U' and V' share no points of W in common. But then W is disconnected, contradicting the assumption that W is connected. Therefore $f(W)$ is connected.

Corollary 1:   If $f: W \to \mathbb{R}$ is continuous and W is a closed interval, then $f(W)$ either consists of one point or is a closed interval.

PROOF: Since f is continuous and W is both compact and connected, $f(W)$ is both compact and connected. By Corollary 2 of Proposition 34, $f(W)$ must be either a one point subset or a closed interval.

Corollary 2 (*The Intermediate Value Theorem*):   If $f: W \to \mathbb{R}$ is continuous and W is connected, then if a and b are any two elements of $f(W)$ with $a < b$, and if $a < c < b$, then there is at least one $w \in W$ such that $f(w) = c$. (An illustration of this corollary relative to the graph of f is given in Figure 13.)

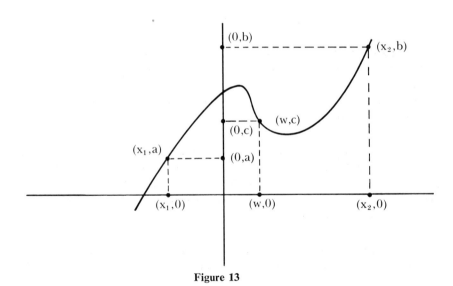

Figure 13

PROOF: Since W is connected, $f(W)$ is connected. Then by Proposition 34, $[a, b] \subseteq f(W)$. Since $c \in [a, b] \subseteq f(W)$, there must be $w \in W$ such that $f(w) = c$.

We can prove Corollary 2 without appealing to connectedness.

Alternate proof of Corollary 2:   Set $A = \{x \mid x \in [a, b], f(x) < c\}$. Since $a \in A$, A is non-empty. Moreover, b is an upper bound for A; hence A has a least upper bound u. There are sequences $\{s_n\}$, $n \in \mathbb{N}$, and $\{t_n\}$, $n \in \mathbb{N}$, of points of $[a, b]$ such that $s_n < u$ and $u < t_n$ for all $n \in \mathbb{N}$ (since we are assuming c is neither $f(a)$ or $f(b)$) with $s_n \to u$ and $t_n \to u$. Since f is continuous $f(s_n) \to f(u)$ and $f(t_n) \to f(u)$. Since $f(s_n) \leqslant c$ for all $n \in \mathbb{N}$ $f(u) \leqslant c$. But since $f(t_n) \geqslant c$ for all $n \in \mathbb{N}$, $f(u) \geqslant c$. Therefore $f(u) = c$.

As an application of the Intermediate Value Theorem, we give the following classical result.

Proposition 36: Any positive real number has at least one real nth root for any positive integer n.

PROOF: Consider $f(x) = x^n$, where n is an positive integer, and let a be any positive real number. Then f is continuous (we will prove this later), $f(0) = 0 < a$, and there is $b \in \mathbb{R}$ such that $f(b) = b^n > a$. Since f is defined for all of $\mathbb{R}$ and f is continuous (to be proved rigorously later), $f(\mathbb{R})$ is connected. Since $0 < a < b^n$, there is some $x \in \mathbb{R}$ such that $f(x) = x^n = a$ by Corollary 2 of Proposition 35. Therefore x is an nth root of a.

We now use the Intermediate Value Theorem to prove the following important result.

Proposition 37: Suppose f is defined and strictly increasing (strictly decreasing) on the open interval (a, b) (that is, if $x, x' \in (a, b)$ with $x < x'$, then $f(x) < f(x')$). Then

a) f is one-one; hence $f^{-1}: f((a, b)) \to \mathbb{R}$ is a function;
b) f^{-1} is strictly increasing (decreasing); and
c) if f is continuous, then f^{-1} is also continuous.

PROOF

(a) Suppose $f(x) = f(x')$. If $x < x'$, then $f(x) < f(x')$, and if $x' < x$, then $f(x') < f(x)$. It follows then that $x = x'$; hence f is one-one. Since f is one-one and onto $f((a, b))$, $f^{-1}: f((a, b)) \to \mathbb{R}$ is a function.

(b) If f^{-1} is not strictly increasing, then we have $y, y' \in f((a, b))$ such that $y < y'$ but $f^{-1}(y') < f^{-1}(y)$. Since y and y' are in the image of f, there are x and x' in (a, b) such that $f(x) = y$ and $f(x') = y'$. Therefore $y < y'$ and $f^{-1}(y') < f^{-1}(y)$ implies $f(x) < f(x')$ for $f^{-1}(f(x')) = x' < x = f^{-1}(f(x))$, a contradiction to f being strictly increasing. Therefore f^{-1} is strictly increasing.

(c) Let $x \in (a, b)$ and choose $p > 0$; we may choose p sufficiently small such that $x - p$ and $x + p$ are both in (a, b) (Fig. 14). Since $x - p < x < x + p, f(x - p) < f(x) < f(x + p)$. Using the strictly increasing property of f, we can prove $f((x - p, x + p)) \subseteq (f(x - p), f(x + p))$. Using the Intermediate Value Theorem and the strictly increasing property of f, we

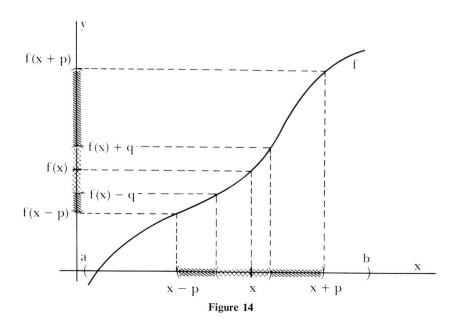

Figure 14

can prove

$$f((x - p, x + p)) = (f(x - p), f(x + p)).$$

Choose $q > 0$ such that $(f(x) - q, f(x) + q) \subseteq (f(x - p), f(x + p))$. Then if y is in $(f(x) - q, f(x) + q)$, that is, $|y - f(x)| < q$, then $f^{-1}(y) \in (x - p, x + p)$, that is $|f^{-1}(y) - f^{-1}(f(x))| = |f^{-1}(y) - x| < p$. Therefore f^{-1} is continuous.

EXERCISES

1. In Example 26, prove that given any $p > 0$ and any $q > 0$, there is $x \in W$ such that $f(N(x, q)) \not\subseteq N(f(x), p)$.

2. In the proof of Proposition 31, prove that

$$f(N(w, q) \cap W) \subseteq N(f(w), p)$$

for each $w \in W$.

3. Use Proposition 35 to prove that any open interval, open half-line, and $\mathbb{R}$ are connected. For example, if (a, b) is any open interval, find a continuous function from $(0, 1)$ onto (a, b).

4. Prove Corollary 1 of Proposition 34.

5. Prove that the only subsets of $\mathbb{R}$ which are both closed and open are $\mathbb{R}$ and $\varnothing$.

6. Prove that W is disconnected if and only if there is a continuous function $f: W \to \mathbb{R}$ such that the image of W is $\{0, 1\}$.

7. Prove or disprove: A subset W of $\mathbb{R}$ is compact if and only if each non-empty subset of W has both a lub and glb which is an element of W.

8. Prove that if $f:(a, b) \to \mathbb{R}$ is strictly increasing and continuous, then $f((a, b))$ is an open interval, open half-line, or all of $\mathbb{R}$.

9. Suppose $f:[a, b] \to \mathbb{R}$ is continuous with $f(a) f(b) < 0$. Prove there is $w \in (a, b)$ such that $f(w) = 0$.

10. Prove that if $f:[0, 1] \to [0, 1]$ is continuous, then f has a fixed point, that is, there is $w \in [0, 1]$ such that $f(w) = w$. Hint: Show the function g defined by $g(x) = f(x) - x$ is 0 for some $x \in [0, 1]$.

11. Prove Proposition 37 for the case that f is strictly decreasing.

4

SEQUENCES AND SERIES

4.1 MORE PROPERTIES OF SEQUENCES

OPERATIONS WITH SEQUENCES

It is possible to define operations on the set of sequences of real numbers in the following way.

Definition 1: *Suppose* $s = \{s_n\}$, $n \in \mathbb{N}$, *and* $t = \{t_n\}$, $n \in \mathbb{N}$, *are sequences of real numbers and r is any real number. Then we define*

$s + t$, *the **sum** of s and t, to be* $\{s_n + t_n\}$, $n \in \mathbb{N}$,
st, *the **product** of s and t, to be* $\{s_n t_n\}$, $n \in \mathbb{N}$,
rs, *the **scalar multiple** of s by r, to be* $\{rs_n\}$, $n \in \mathbb{N}$, *and*
$s^{-1} = \{s_n^{-1}\}$, $n \in \mathbb{N}$, *provided that no* $s_n = 0$.

We are primarily interested in the relationship between the convergence of s and t and the convergence of the sum, product, and scalar multiples of s and t. The basic results are given in Proposition 2. We first prove a lemma.

Proposition 1: A sequence s converges to L if and only if the sequence t defined by $t_n = s_n - L$ converges to 0.

PROOF: If t converges to 0, then for any $p > 0$, there is an integer M such that if $n > M$, $|t_n - 0| = |s_n - L| < p$. Therefore $n > M$ implies $|s_n - L| < p$; consequently, $s_n \to L$. On the other hand, if $s_n \to L$, then

66

for any $p > 0$, there is an integer M such that $n > M$ implies $|s_n - L| = |t_n - 0| < p$; therefore $t_n \to 0$.

Proposition 2: Suppose $s_n \to L$ and $t_n \to L'$. Then

a) $s_n + t_n \to L + L'$,
b) $rs_n \to rL$,
c) $s_n t_n \to LL'$, and
d) $s_n^{-1} \to L^{-1}$, provided $L \neq 0$ and s^{-1} is defined.

PROOF

(a) Let $p > 0$. Then there is an integer M_1 such that $n > M_1$ implies $|s_n - L| < p/2$. There is also an integer M_2 such that $n > M_2$ implies $|t_n - L'| < p/2$. Let M be the greater of M_1 and M_2. Then if $n > M$, $|(s_n + t_n) - (L + L')| \leqslant |s_n - L| + |t_n - L'| < p/2 + p/2 = p$. Therefore $s_n + t_n \to L + L'$.

The proof of (b) is left as an exercise.

(c) A simple computation shows that $s_n t_n - LL' = s_n(t_n - L') + (s_n - L)L'$. Since $s_n \to L$ and $t_n \to L'$, $t_n - L'$ and $s_n - L$ both converge to 0. But then $s_n t_n - LL'$ converges to 0; hence $s_n t_n \to LL'$.

(d) There is an integer M_1 such that $n > M_1$ implies $|s_n - L| < |L|/2$. It follows then that $|L|/2 < |s_n|$ if $n > M_1$. Let $p > 0$. Then there is an integer M_2 such that if $n > M_2$, $|s_n - L| < p |L|^2/2$. Let M be the greater of M_1 and M_2. Then if $n > M$,

$$|1/s_n - 1/L| = |(s_n - L)/s_n L| < (2/|L|^2) |s_n - L| < p.$$

Therefore $s_n^{-1} \to 1/L$.

OPERATIONS WITH FUNCTIONS

Proposition 2 has immediate consequences concerning the continuity of functions. We first define algebraic operations for functions.

Definition 2: *Suppose f and g are both functions from W into* $\mathbb{R}$. *Then we define*

a) $f + g: W \to \mathbb{R}$ *by* $(f + g)(w) = f(w) + g(w)$ *for each* $w \in W$.
b) $fg: W \to \mathbb{R}$ *by* $(fg)(w) = f(w)g(w)$ *for each* $w \in W$.
c) $rf: W \to \mathbb{R}$ *by* $(rf)(w) = rf(w)$ *for each* $w \in W$, *where* r *is any real number.*
d) $1/f: W \to \mathbb{R}$ *by* $(1/f)(w) = 1/f(w)$ *for each* $w \in W$ *for which* $f(w) \neq 0$.

Example 1: Let $f(x) = x^3$ and $g(x) = x + 4$, both f and g defined on all of $\mathbb{R}$. Then $(f + g)(x) = x^3 + x + 4$, $(fg)(x) = x^3(x + 4)$, $(7f)(x) = 7x^3$, and $(1/g)(x) = 1/(x + 4)$ for $x \neq -4$.

Proposition 3: If f and g are both functions from W into $\mathbb{R}$ which are continous at $a \in W$ then $f + g$, fg, rf for any real number r, and $1/f$ provided $f(a) \neq 0$ are continuous at a.

PROOF: We will prove $f + g$ is continuous at a. The technique for proving continuity of the other functions is similar. Suppose s_n is a sequence in W which converges to a in W. Since f and g are both continuous, $f(s_n) \to f(a)$ and $g(s_n) \to g(a)$. Then $(f + g)(s_n) = f(s_n) + g(s_n)$ converges to $f(a) + g(a) = (f + g)(a)$ by Proposition 2. Therefore $f + g$ is continuous at a by Proposition 19 of Chapter 3.

Corollary: If f and g are both continuous functions from W into $\mathbb{R}$, then $f + g$, fg, rf for any real number r, and $1/f$ whenever defined are continuous.

Example 2: The function $i: \mathbb{R} \to \mathbb{R}$ defined by $i(x) = x$ is continuous. Therefore the function $i^2(x) = x^2$ is continuous. By induction, $i^n(x) = x^n$ is continuous for any positive integer n. Therefore $(ri^n)(x) = rx^n$ is continuous for any real number r. From this it follows readily that any polynomial function f, that is, a function of the form

$$f(x) = a_n x^n + a_{n-1} x^{n-1} + \ldots + a_1 x + a_0, \quad a_i \in \mathbb{R}, \quad i = 1, \ldots, n,$$

is continuous.

Example 3: The function f defined by $f(x) = x^{3/2}$ for x in $(0,1)$ is continuous since it is the composition of the function g defined by $g(x) = x^3$, which is continuous, with the function h defined by $h(x) = x^{1/2}$, which is also continuous. We know that h is continuous since $h = g^{-1}$, where $g(x) = x^2$, $x \geq 0$. Since g is continuous and strictly increasing, g^{-1} is continuous by Proposition 37 of the last chapter.

Proposition 2 of this chapter and Proposition 20 of Chapter 3 are very useful in proving the continuity of many functions.

SUBSEQUENCES

Now consider the sequence defined by $s_n = (-1)^n$, that is,

$$-1, 1, -1, 1, -1, \ldots . \tag{1}$$

This sequence does not converge; for if it converged, it would have to have either -1 or 1 as its limit, but it does not stay in the 1-neighborhoods of either -1 or 1. (We also see (1) cannot converge since it is not a Cauchy sequence.) Nevertheless, there is a *subsequence* of (1) which converges to 1; for if we took every even term from (1), we would have a sequence whose terms are all 1. Taking every even term of (1) does not really give a sequence since a sequence is by definition a function from $\mathbb{N}$ into $\mathbb{R}$. We can correctly obtain the result we want (that is, having the even terms of (1) in the proper order) and have a genuine sequence in the following way: Define $k: \mathbb{N} \to \mathbb{N}$ by $k(n) = 2n$. Set $s_{k_n} = (s \circ k)(n) = s_{2n}$. Then $\{s_{k_n}\}$, $n \in \mathbb{N}$, really is a sequence, and it does give us the even terms of s in the same order that they appear in (1). The sequence $\{s_{k_n}\}$, $n \in \mathbb{N}$, is an example of a *subsequence* of $\{s_n\}$, $n \in \mathbb{N}$. The general definition of the useful notion of a subsequence is given in the following definition.

Definition 3: *Let s be a sequence of real numbers. Suppose $k: \mathbb{N} \to \mathbb{N}$ is such that*

$$k(n) < k(m), \quad \text{whenever } n < m. \tag{2}$$

*Then the function $s \circ k: \mathbb{N} \to \mathbb{R}$ is said to be a **subsequence** of s.*

A subsequence is, of course, a sequence in its own right.

Since the function $k: \mathbb{N} \to \mathbb{N}$ defined by $k(n) = 2n$ has property (2), $s \circ k$ defined in the paragraph preceding Definition 3 is indeed a subsequence of (1). We see in the preceding example that a sequence which does not converge may, nevertheless, have a convergent subsequence. The next example gives another example of such a situation.

Example 4: Let s be the sequence defined by $s_n = 1/n$ if n is even and $s_n = 1$ if n is odd. Since there are even integers which are arbitrarily large, s does not converge. But it does have a subsequence which converges; in fact, it has infinitely many convergent subsequences, each of which converges to one of two numbers. Define $k: \mathbb{N} \to \mathbb{N}$ by $k(n) = 2n$ and $k': \mathbb{N} \to \mathbb{N}$ by $k'(n) = 2n - 1$. Then $s \circ k$ and $s \circ k'$ are both subsequences of s and $s_{k_n} \to 0$, while $s_{k'_n} \to 1$.

Although a sequence which does not converge may (or may not) have a convergent subsequence, no sequence which converges can have a subsequence which does not converge as we see from the following proposition.

Proposition 4: A sequence s converges to a limit L if and only if every subsequence of s converges to L.

PROOF: Since $s = s \circ k$, where $k(n) = n$ for each positive integer n, s is a subsequence of itself. If every subsequence of s converges to L, then s must converge to L. Assume now that s converges to L and $s \circ k$ is a subsequence of s. Let $p > 0$. Since $s_n \to L$, there is an integer M such that $n > M$ implies $|s_n - L| < p$. Since $k: \mathbb{N} \to \mathbb{N}$ has property (2), there is an integer M' such that $n > M'$ implies $k(n) > M$. Therefore if $n > M'$, we have $|s_{k_n} - L| < p$ since $k_n > M$. Therefore $s_{k_n} \to L$, and the proposition is proved.

LIMIT POINTS

The fact that a sequence can have a subsequence which converges, or even several subsequences which converge to different limits, even when the original sequence does not converge, inspires the following definition.

Definition 4: *A real number a is said to be a **limit point** of a sequence s if s has a subsequence which converges to a.*

Proposition 4 tells us that a limit of a sequence is a limit point of a sequence. The limit points of the sequence in Example 4 are 0 and 1. Limit points of sequences and limit points of sets are related to one another in the following way.

Proposition 5: A number a is a limit point of a set W if and only if there is a sequence of points of $W \sim \{a\}$ which converges to a.

PROOF: If there is a sequence of points of $W \sim \{a\}$ which converges to a, then every open set which contains a must contain a least one point of W other than a; hence a is a limit point of W. Suppose that a is a limit point of W and set $p_n = 1/n$. For each p_n, select s_n contained in $N(a, p_n) \sim \{a\}$ (such a selection is possible because a is a limit point of W). Then $\{s_n\}$, $n \in \mathbb{N}$, is a sequence of points of $W \sim \{a\}$ which converges to a.

The next proposition gives an important criterion for compactness in terms of limit points.

Proposition 6: A set W is compact if and only if every sequence of points of W has a limit point in W.

PROOF: First, suppose that W is compact and s is a sequence of points of W. Then s has a monotone increasing or decreasing subsequence (the proof of this fact is left to the reader). Since W is bounded, this monotone subsequence is a bounded monotone sequence and hence converges (Exercise 4, Section 3.2). Since W is closed, the limit of this subsequence, a limit point of s, is contained in W (Exercise 2 of Section 3.4).

Next, suppose that every sequence in W has a limit point in W. Suppose s is a sequence of points of W which converges to L. Since L is the only limit point of s (Proposition 4), L must be in W; therefore W is closed by Exercise 2 of Section 3.4. We now show that W is also bounded and, hence, is compact since it is then closed and bounded. If W is not bounded, then W has either no upper bound or no lower bound. Assume the former; the proof for the latter case is quite similar. Then we can find $s_1 \in W$ such that $s_1 > 1$. Let n_1 be the least integer greater than s_1. Choose $s_2 \in W$ with $s_2 > n_1$. Let n_2 be the least integer greater than s_2 and select s_3 from W with $s_3 > n_2$. Continuing in like fashion we obtain a sequence $s_1, s_2, s_3, \ldots$ of points of W which has no limit points (the proof of this fact is left to the reader), but this contradicts the assumption that every sequence in W has a limit point. Therefore W must be bounded and, hence, compact.

We now prove the converse of Proposition 27 of Chapter 3.

Proposition 7:   A set W is compact if every infinite subset of W has a limit point.

PROOF: Let s be a sequence in W. If s takes but a finite number of values, then at least one of these values must be a limit point of s (the proof is left to the reader); hence s has a limit point. If s takes infinitely many values, then the set of s_n is an infinite subset of W and hence has a limit point L in W. We now prove that L is also a limit point of s.

Since L is a limit point of the set of s_n, each open set which contains L contains infinitely many s_n. For if there were some open set U which contained finitely many s_n, we could find an open set containing L which contained none of the s_n as follows: Suppose U is an open set which contains L and only $s_{n_1}, \ldots, s_{n_k}$. For each $s_{n_j}, j = 1, \ldots, k$, we can find p_j such that (i) $N(L, p_j) \subseteq U$ and (ii) $N(s_{n_j}, p_j)$ contains no points in common with $N(L, p_j)$. Let p be the minimum of the p_j. Then $N(L, p)$ is an open set which is contained in U and contains none of the $s_{n_j}, j = 1, \ldots, k$. Hence $N(L, p)$ is an open set which contains L but contains none of the s_n since the only s_n contained in U are the s_{n_j}. This contradicts the assumption that L is a limit point of the set of s_n. Therefore each open set which contains L contains infinitely many s_n. Set $p_n = 1/n$. We now define $k: \mathbb{N} \to \mathbb{N}$ as follows: Let $k(1)$ be the least integer n for which $s_n \in N(L, 1)$. Assume $k(n-1)$ has been defined. Define $k(n)$ to be the least integer n greater than $k(n-1)$ for which $s_n \in N(L, p_n)$. Then $s \circ k$ is a subsequence of s and $s \circ k$ converges to L; hence L is a limit point of s. We have therefore shown that any sequence of points of W has a limit point in W; hence the proof of Proposition 7 is complete.

Example 5:   We see that the open interval $(0, 1)$ is not compact since the sequence $s_n = 1/n$ in $(0, 1)$ does not have a limit point in $(0, 1)$. (The only limit point of s is 0, which is not a point of $(0, 1)$.)

EXERCISES

1. Prove (b) of Proposition 2.

2. Suppose f and g are continuous functions from W into $\mathbb{R}$ and r is any real number. Prove that $fg: W \to \mathbb{R}$ and $rf: W \to \mathbb{R}$ are also continuous. Prove that $1/f$ is continuous where it is defined. Hint: $(1/f) = (1/i) \circ f$, where $i(x) = x$.

3. Prove the existence of the monotone sequence required n the proof of Proposition 6. In the second part of that same proof, show that the sequence constructed has no limit point. This may be done, for example, by showing that any subsequence is not bounded and, hence, cannot converge.

4. Prove that any sequence which takes but a finite number of values must have a limit point.

5. Prove that every bounded sequence has a convergent subsequence.

6. Prove that a set is closed if and only if it contains all of its limit points.

7. Let t be a subsequence of s and let t' be a subsequence of t. Prove that t' is also a subsequence of s.

8. Prove that a sequence s does not converge to L if and only if s has some subsequence no subsequence of which converges to L.

9. Prove that a sequence s is a Cauchy sequence if and only if $\lim_{j \to \infty} |s_{n_{j+1}} - s_{n_j}| = 0$ for every subsequence $\{s_{n_j}\}$ of s.

10. Suppose s is a sequence; let A be the set of limit points of s. If A is non-empty and bounded above, we set

$$\limsup_{n \to \infty} s_n = \text{lub } A;$$

if A is non-empty and bounded below, we set

$$\liminf_{n \to \infty} s_n = \text{glb } A.$$

We let $\limsup\limits_{n\to\infty} s_n = \infty$ if s has a subsequence which converges to ∞, and $\liminf\limits_{n\to\infty} s_n = -\infty$ if s has a subsequence which converges to $-\infty$.

    a) Find $\liminf\limits_{n\to\infty} s_n$ and $\limsup\limits_{n\to\infty} s_n$ for each of the following:
        (i) $s_n = (-1)^n$; (ii) $s_n = 1$, n even, $s_n = n$, n odd;
        (iii) $s_n = 1/n$.
    b) Suppose s and t are sequences such that $s_n \leqslant t_n$ for $n > M$, M some integer. Prove that $\liminf\limits_{n\to\infty} s_n \leqslant \liminf\limits_{n\to\infty} t_n$ and $\limsup\limits_{n\to\infty} s_n \leqslant \limsup\limits_{n\to\infty} t_n$.
    c) Prove that $s_n \to L$, L finite, if and only if $\liminf\limits_{n\to\infty} s_n = \limsup\limits_{n\to\infty} s_n = L$.

    11. Suppose s and t are bounded sequences of real numbers. Prove that

$$\limsup_{n\to\infty}(s_n + t_n) \leqslant \limsup_{n\to\infty} s_n + \limsup_{u\to\infty} t_n$$

and

$$\liminf_{n\to\infty}(s_n + t_n) \geqslant \liminf_{n\to\infty} s_n + \liminf_{n\to\infty} t_n.$$

Provide examples to show that equality need not hold.

4.2 SERIES

    We assume that the reader has encountered the standard summation notation earlier in his mathematical studies. To refresh his memory, if need be, we review this notation briefly now.

    The greek capital letter sigma, $\sum$, is used to indicate a sum is to be taken. Above, below, and following the $\sum$ are "instructions" as to what sum is to be taken. The sums which are of particular interest to us in this section have the form

$$\sum_{i=q}^{n} a_i, \quad q \text{ and } n \text{ integers}, \quad q \leqslant n. \tag{3}$$

The notation in (3) is but a shorthand way of writing $a_q + a_{q+1} + \ldots + a_n$. We call i the *index of summation;* we substitute for i in the *summand* a_i the values called for in the instructions for summation. The $i = q$ below and n above the $\sum$ indicate that all integer values of i between q and n, inclusive, are to be used. (Of course the letter i need not always be the index of summation; j, k, and so forth would also work. But we would have to be sure that the index of summation in the instructions of summation and in the summand matched appropriately according to the sum we wanted to denote.)

Definition 5:   *Let* $\{a_n\}$, $n \in \mathbb{N}$, *be any sequence, and set*

$$t_n = \sum_{i=1}^{n} a_i. \tag{4}$$

Then the sequence $\{t_n\}$, $n \in \mathbb{N}$, *is said to be the* **series based on** $\{a_n\}$, $n \in \mathbb{N}$, *and* a_n *is said to be the* **nth term** *of this series. We call* t_n *the* **nth partial sum** *of this series. We say that the series* **converges to,** *or has* **sum,** L *if* $t_n \to L$. *If a series does not converge, we say that it* **diverges.** *The series based on* $\{a_n\}$, $n \in \mathbb{N}$, *is generally denoted by* $\sum_{n=1}^{\infty} a_n$;* *for simplicity, we will often denote this series by writing* $\sum a_n$ *provided that no ambiguity can result.*

Example 6:   Let $a_n = (1/2)^n$. Then $\sum a_n$ has the sequence of partial sums $\{t_n\}$, $n \in \mathbb{N}$, where $t_n = \sum_{i=1}^{n} (1/2)^i$. $\sum a_n$ is a *geometric series* with *first term* 1/2 and ratio 1/2. By the standard formula for the sum of a geometric series, we find that this series converges to $(1/2)/(1 - 1/2) = 1$ (cf. Proposition 11 below).

Geometric series play such an important role in any discussion of series that we will soon define such series rigorously and discuss conditions under which they converge. Preliminary to this, however, other material must be introduced.

The proposition which follows is simply a statement of Proposition 13 of Chapter 3 as applied to series.

Proposition 8:   A series $\sum a_i$ converges if and only if given any $p > 0$, there is an integer M such that if n and m are integers greater than M and $m \leqslant n$, then

$$\left| \sum_{i=m}^{n} a_i \right| < p. \tag{5}$$

(Note that the left side of (5) is nothing but $|t_n - t_{m-1}|$.)

Proposition 9:   In order for the series $\sum a_i$ to converge, it is necessary (but not sufficient) for $a_n \to 0$.

---

* In some instances, $\sum_{n=1}^{\infty} a_n$ is also used to denote the sum of the series.

PROOF: If $\sum a_i$ converges, then if we take any $p > 0$, there is an integer M such that for $n > M$, $\left| \sum_{i=n}^{n} a_i \right| = |a_n| < p$. From this it follows that $a_n \to 0$.

The following example shows that $a_n \to 0$ is not sufficient to insure convergence of $\sum a_i$.

Example 7: Let $a_n = 1/n$. Then $a_n \to 0$. However, $\sum a_i$ diverges. There are several ways to prove this latter fact, perhaps the simplest and most classical being the following: $t_1 = a_1 = 1$. $t_2 = 1 + 1/2 = 3/2$. $t_4 = (1 + 1/2) + (1/3 + 1/4) > t_2 + 1/2 = 2$. $t_8 = t_4 + (1/5 + 1/6 + 1/7 + 1/8) > t_4 + 4(1/8) = t_4 + 1/2 = 5/2$. We can proceed in like fashion to prove that $t_{2^n} > (n + 1)/2$; or alternatively,

$$t_{2^{n+1}} - t_{2^n} = \sum_{j=2^n+1}^{2^{n+1}} \frac{1}{j} > \sum_{j=2^n+1}^{2^{n+1}} \frac{1}{2^{n+1}} = (2^{n+1} - 2^n) \frac{1}{2^{n+1}} = \frac{1}{2}$$

so $\{t_n\}$, $n \in \mathbb{N}$, is not a Cauchy sequence. Hence t_n cannot converge to a finite limit.

Proposition 10: $r^n \to 0$ if and only if $|r| < 1$.

PROOF: Clearly $r^n \to 0$ if $r = 0$. Assume $0 < r < 1$. Then $\{r^n\}$, $n \in \mathbb{N}$, is a bounded monotone decreasing sequence which has 0 as a lower bound. Therefore $r^n \to L$ for some number $L \geq 0$. If we multiply $\{r^n\}$, $n \in \mathbb{N}$, by r, then we obtain the sequence $\{r^{n+1}\}$, $n \in \mathbb{N}$, which converges to rL. But $\{r^{n+1}\}$, $n \in \mathbb{N}$, must converge to the same limit as the original sequence (why?). Consequently, $L = rL$, or

$$L(1 - r) = 0. \tag{6}$$

Since $1 - r \neq 0$, it follows from (6) that $L = 0$.

Suppose $-1 < r < 0$. Then $|r^n| = |r^n| \to 0$. Therefore $r^n \to 0$ (see Example 8 of Chapter 3).

If $1 < r$ and $r^n \to L$, then $1 \leq L$. But, by the same reasoning as the first part of the proof, $r^{n+1} \to rL = L$; hence $L = 0 < 1$, a contradiction. Consequently, $\{r^n\}$, $n \in \mathbb{N}$, does not converge for $1 < r$. We leave it for the reader to prove that $\{r^n\}$, $n \in \mathbb{N}$, does not converge if $r < -1$. Clearly too,* $r^n \nrightarrow 0$ for $r = 1$ or $r = -1$. This completes the proof of Proposition 10.

GEOMETRIC SERIES

We now return to geometric series.

Definition 6: *Let a and r be any non-zero real numbers. The sequence defined by* $s_n = ar^{n-1}$ *is said to be a* **geometric progression with first term** *a*

---

* $r^n \nrightarrow 0$ means that r^n does not converge to 0.

and **ratio** r. *The series based on this geometric progression is said to be a* **geometric series** *with* **first term** *a and* **ratio** *r.*

We generally make a minor change in the index of summation and denote such a geometric series by

$$\sum_{i=0}^{\infty} ar^i, \tag{7}$$

or simply by $\sum ar^i$ when no ambiguity might result.

Proposition 11: The geometric series (7) converges if and only if $|r| < 1$. If (7) converges, it converges to $a/(1 - r)$.

PROOF: Since $a \neq 0$, it follows that (7) does not converge if $r = 1$ since in that case ar^n does not converge to 0. Assume then that $r \neq 1$. Then the nth partial sum t_n of (7) is given by

$$t_n = a + ar + \ldots + ar^n. \tag{8}$$

Multiplying (8) by r we obtain

$$rt_n = ar + ar^2 + \ldots + ar^n + ar^{n+1}. \tag{9}$$

Subtracting (9) from (8), we find $t_n - rt_n = a - ar^{n+1}$. Solving for t_n, we find

$$t_n = \frac{a(1 - r^{n+1})}{1 - r}. \tag{10}$$

From Proposition 10, $r^{n+1} \to 0$ if and only if $|r| < 1$; it follows then that $\{t_n\}$, $n \in \mathbb{N}$, converges if and only if $|r| < 1$; and if $\{t_n\}$, $n \in \mathbb{N}$, converges, it converges to $a/(1 - r)$.

TESTS FOR CONVERGENCE

We now have a large family of convergent series. We can use this family of convergent geometric series to help find other convergent series in accordance with the following proposition.

Proposition 12 (*The Comparison Test*)

a) Suppose $\sum a_i$ is a convergent series all of whose terms except finitely many are non-negative, and suppose $\sum b_i$ is a series such that $0 \leqslant b_i \leqslant a_i$

for i sufficiently large (that is, for $i \geqslant M$ for some integer M). Then $\sum b_i$ converges.

b) Suppose $\sum a_i$ is a divergent series all but finitely many terms o which are non-negative, and suppose $\sum b_i$ is a series such that $a_i \leqslant b_i$ if $i \geqslant M$ for some integer M. Then $\sum b_i$ diverges.

PROOF: We prove (a) and leave the proof of (b) as an exercise. Let M be such that $i \geqslant M$ implies $0 \leqslant b_i \leqslant a_i$ and set

$$w_n = \sum_{i=M}^{n} b_i \quad \text{for} \quad n \geqslant M.$$

Then the w_n form a monotone increasing sequence which is bounded above by $\sum_{i=M}^{n} a_i$. Therefore $w_n \to L$ for some number L. Since the nth partial sum of $\sum b_i$ for $n \geqslant M$ is given by

$$(b_1 + b_2 + \ldots + b_{M-1}) + w_n,$$

it follows that $\sum b_i$ converges to $\left(\sum_{i=1}^{M-1} b_i \right) + L$.

Example 8: Consider the series $\sum_{i=0}^{\infty} 1/i!$, where $i!$ is defined to be 1 if $i = 0$ and the product of all positive integers less than or equal to i if $i \geqslant 1$. Thus $i! = i(i-1)(i-2) \ldots 2 \cdot 1$ for $i \geqslant 1$. Then $0 \leqslant 1/i! \leqslant (1/2)^i$ for $i \geqslant 4$. Since $\sum (1/2)^i$ converges, $\sum 1/i!$ also converges.

Example 9: The decimal representation

$$e_n e_{n-1} \ldots e_0 \cdot d_1 d_2 d_3 d_4 \ldots$$

is a shorthand way of writing

$$e_n \cdot 10^n + \ldots + e_0 \cdot 10^0 + d_1 \cdot 10^{-1} + d_2 \cdot 10^{-2} \ldots, \tag{11}$$

where each e_i and d_i is an integer between 0 and 9, inclusive. Each term of (11) is smaller than the corresponding term of the geometric series with first term 10^{n+1} and ratio $1/10$. Since this latter series converges, (11) also converges by the Comparison Test. We have thus shown that the usual decimal representation of a real number represents a convergent series.

Proposition 13 (*The Ratio Test*): Suppose $\sum a_i$ is a series all of whose terms are positive except for finitely many, and set

$$s_n = a_{n+1}/a_n.$$

If there is $r < 1$ such that for some integer M, $n > M$ implies $s_n < r$, then $\sum a_i$ converges. If there is $r > 1$ such that for some integer M, $n > M$ implies $s_n > r$, then $\sum a_i$ diverges.

PROOF: If there is $r < 1$ such that $s_n < r$ for $n > M$, then convergence follows from comparison with the geometric series $\sum a_1 r^i$. If there is $r > 1$ such that $n > M$ implies $a_{n+1}/a_n > r > 1$, then $a_{n+1} > a_n$; hence $a_n \nrightarrow 0$. Therefore $\sum a_i$ diverges.

The following example shows that the limit of a_{n+1}/a_n does not always exist even when $\sum a_n$ is convergent.

Example 10:   Consider $\sum\limits_{n=1}^{\infty} a_n$, where $a_n = (1/3)^n$ if n is odd and $a_n = (1/2)^n$ if n is even. Then if n is odd, $a_{n+1}/a_n = (1/2)^{n+1}/(1/3)^n = (3/2)^n(1/2)$, while if n is even, $a_{n+1}/a_n = (1/3)(2/3)^n$. It follows then that a_{n+1}/a_n has no limit as n approaches infinity. By comparison with the geometric series $\sum\limits_{n=1}^{\infty} (1/2)^n$, however, we see that the given series converges.

Still another test for convergence which can be proved using the Comparison Test and geometric series is the *Root Test*.

Proposition 14 (*The Root Test*):   Assume that $\sum a_i$ is a series all but finitely many terms of which are non-negative. Then if there is k with $0 \leqslant k < 1$ such that for some integer M, $n > M$ implies $a_n^{1/n} \leqslant k$, then $\sum a_i$ converges. If there is $k > 1$ such that $n > M$ for some integer M implies $a_n^{1/n} \geqslant k$, then $\sum a_i$ diverges.

PROOF: $a_n^{1/n} \leqslant k$ if and only if $a_n \leqslant k^n$. We can therefore compare $\sum a_i$ with the geometric series $\sum k^i$.

The tests developed thus far for series have two drawbacks. For the most part they are stated for series all but finitely many of whose terms are non-negative. Since there are many series with an infinite number of negative terms, our tests are not as general as we might want. Secondly, there are a great many series which satisfy the condition that all but finitely many of their terms are non-negative, yet for which the preceding tests do not work. In some of these latter instances, we cannot find an apt series for comparison, or the limit L of Proposition 13 is 1, or some other unfortunate accident keeps us from gaining much useful information about the series. Although we will find other tests to help us to determine if a series converges, the fact remains that many series are extremely difficult to work with. This we can do little about. We will soon investigate series with infinitely many negative terms, however, and thus some of the restrictions imposed on the tests in Propositions 12, 13, and 14 will be modified or removed.

Note too that we are generally not so much interested in what a series converges to as whether or not it converges. A good computer might sum a series to any degree of accuracy we want, but it is useless to program a computer to sum a series unless we are certain that the series in question converges.

Much of theory in this chapter is also being developed with an eye toward investigating power series and series expansion of functions in a later portion of this text. At that time we will find we have most of the machinery needed to solve the problems we will be interested in.

Before proceeding to discuss arbitrary series, we define two algebraic operations for series.

Definition 7: *Given the series $\sum a_i$ and $\sum b_i$, we define the* **sum** *of these series to be $\sum (a_i + b_i)$. We denote the sum by $\sum a_i + \sum b_i$. If r is any real number, we define $r(\sum a_i) = \sum ra_i$.*

There is also an "appropriate" way to multiply $\sum a_i$ and $\sum b_i$ (the *product* is not $\sum a_i b_i$) which we will discuss in the next section. The following proposition follows easily from what we already know about addition and scalar multiplication of sequences.

Proposition 15: If $\sum a_i$ converges to L and $\sum b_i$ converges to L', then $\sum a_i + \sum b_i$ converges to $L + L'$. Moreover, for any real number r, $r(\sum a_i)$ converges to rL.

The proof is left to the reader.

EXERCISES

1. Prove Proposition 15.

2. Prove that if $|r| > 1$, then $\{r^n\}$, $n \in \mathbb{N}$, does not converge.

3. Prove (b) of Proposition 12.

4. Using all the tests introduced so far, prove that $\sum 1/n^n$ converges.

5. Prove that series all but finitely many of whose terms are positive converges if and only if its partial sums have an upper bound.

6. a) Find an example of a series $\sum a_n$ which converges, but such that $\sum a_n^2$ fails to converge.

b) Suppose $\sum a_n$ converges. Is it possible then for $\sum na_n$ to diverge?

7. Reference has been made to "series all but finitely many terms of which are positive." Prove that the convergence or divergence of a series does not depend on any finite number of terms of the series. Specifically, prove that $\sum a_i$ converges if and only if $\sum_{i=M}^{\infty} a_i$ converges for any positive integer M.

8. Use any of the tests developed in this section to test the convergence of the following series. Try to find upper and lower bounds for the limit of any series which converges.

a) $\sum 1/(2n)$
b) $\sum 3^n/n^3$
c) $\sum n^3/3^n$

d) $\sum (n-1)/n^2$
e) $\sum 1/(2^{n-1}3^{n-2})$
f) $\sum n^2/n!$
g) $\sum (n^3/n! + n!/n^n)$

9. Suppose $\sum c_n = \sum a_n + \sum b_n$. Prove that if any two of the given series converge, then the third must also converge.

10. Prove that the series $\sum a_n$, each $a_n > 0$, converges if

$$\limsup_{n \to \infty} a_{n+1}/a_n < 1.$$

For the definition of limsup, see Exercise 10, Section 4.1.

4.3 ABSOLUTE AND CONDITIONAL CONVERGENCE

We have already seen that the series $\sum (1/n)$ does not converge. We now look at the series

$$\sum (-1)^{n+1}(1/n) = 1 - 1/2 + 1/3 - 1/4 + \ldots. \tag{12}$$

The series (12) looks as if it might converge since its partial sums are never greater than 1 nor less than 0. That is, the partial sums of (12) are bounded; hence we can be sure that (12) at least has a convergent subsequence. Indeed, straightforward computation with the first few partial sums heightens the suspicion that (12) converges. We have no test yet, however, to determine if (12) does in fact converge since infinitely many terms of (12) are negative.

The next series of propositions enables us to determine that (12) and many series like (12) do in fact converge.

Proposition 16 is really a lemma to help us prove Proposition 17.

Proposition 16: Suppose $\{a_n\}$, $n \in \mathbb{N}$, and $\{b_n\}$, $n \in \mathbb{N}$, are sequences. Define

$$t_n = \sum_{i=1}^{n} a_i, \quad \text{for any positive integer } n,$$

and set $t_0 = 0$. Then if p and q are positive integers such that $p \leqslant q$, then*

$$\sum_{i=p}^{q} a_i b_i = \sum_{i=p}^{q-1} t_i (b_i - b_{i+1}) + t_q b_q - t_{p-1} b_p. \tag{13}$$

PROOF: $\displaystyle\sum_{i=p}^{q} a_i b_i = \sum_{i=p}^{q} (t_i - t_{i-1}) b_i = \sum_{i=p}^{q} t_i b_i - \sum_{i=p}^{q} t_{i-1} b_i = \sum_{i=p}^{q} t_i b_i$

$\displaystyle - \sum_{j=p-1}^{q-1} t_j b_{j+1}$ (by letting $j = i - 1$) $\displaystyle = \sum_{i=p}^{q-1} t_i(b_i - b_{i+1}) + t_q b_q - t_{p-1} b_p$ (by a straightforward rearrangement of the terms of the sums).

Proposition 17: Let $\{a_n\}$, $n \in N$, be a sequence whose series is $\{t_n\}$, $n \in \mathbb{N}$, that is, t_n is the nth partial sum of the a_i. Suppose (i) $\{t_n\}$, $n \in \mathbb{N}$, is bounded and (ii) $\{b_n\}$, $n \in \mathbb{N}$, is a monotone decreasing sequence which converges to 0. Then $\sum a_i b_i$ converges.

PROOF: Since $\{t_n\}$, $n \in \mathbb{N}$, is bounded, we can find a number K such that $|t_n| < K$ for any $n \in \mathbb{N}$. Since $b_n \to 0$, given any $e > 0$, there is an integer M such that $n > M$ implies $|b_n| < e/(2K)$. Suppose p and q are any integers larger than M; we may assume $p \leqslant q$. Then $\left| \displaystyle\sum_{i=p}^{q} a_i b_i \right| =$

$\left| \displaystyle\sum_{i=p}^{q-1} t_i(b_i - b_{i+1}) + t_q b_q - t_{p-1} b_p \right| \leqslant \left| \displaystyle\sum_{i=p}^{q-1} K(b_i - b_{i+1}) + K b_q + K b_p \right| =$

$K \left| \displaystyle\sum_{i=p}^{q-1} (b_i - b_{i+1}) + b_q + b_p \right| = 2K b_p < e$. We now have that $\sum a_i b_i$ converges by Proposition 8.

Corollary: Suppose $\{s_n\}$, $n \in \mathbb{N}$, is a sequence for which $\{|s_n|\}$, $n \in \mathbb{N}$ is monotone decreasing, $s_n \to 0$, and the terms of $\{s_n\}$, $n \in \mathbb{N}$, alternate in sign (assuming 0 to be both positive and negative for the purposes of this proposition). Then $\sum s_i$ converges.

---

* Formula (13) is sometimes called *Abel's transformation*, or the *partial summation formula*. It is the summation analogy of *integration by parts*.

PROOF: The corollary follows directly from Proposition 17 by letting $a_n = (-1)^n$ or $(-1)^{n+1}$ (whichever is appropriate) and $b_n = |s_n|$.

Example 11: The series in (12) converges because $\{(-1)^{n+1}(1/n)\}$, $n \in \mathbb{N}$, fulfills the conditions of the corollary to Proposition 17. Series (12) is a special case of the general series defined in the following definition.

Definition 8: *A series whose terms alternate sign (at least from some term on) is said to be an **alternating series**.*

Restating the corollary to Proposition 17 in terms of alternating series, we have the following proposition.

Proposition 18: An alternating series $\sum a_n$ converges provided $a_n \to 0$ and $|a_n| \geqslant |a_{n+1}|$ for n sufficiently large.

Definition 9: *A series $\sum a_i$ is said to **converge absolutely** if*

$$\sum |a_i| \tag{14}$$

converges. A series which converges, but does not converge absolutely, is said to ***converge conditionally.***

We note that the series (12) converges, but not absolutely, since $\sum (1/i)$ has been shown to diverge. It is not possible, however, for a series to converge absolutely yet not converge as we see from the following proposition.

Proposition 19: If $\sum a_i$ converges absolutely, then $\sum a_i$ converges.

PROOF: The proposition follows at once from Proposition 8 and the fact that $\left| \sum_{i=p}^{q} a_i \right| \leqslant \sum_{i=p}^{q} |a_i|$.

Proposition 19 enables us to use Propositions 12, 13, and 14 to test for convergence of any series. For even though an arbitrary series $\sum a_i$ may contain infinitely many negative terms (and thus not be a positive series itself), its absolute series $\sum |a_i|$ is a series of non-negative terms. We can then use Propositions 12, 13, and 14 to determine if the series converges absolutely; if the series converges absolutely, then according to Proposition 19, the series converges.

Propositions 17 and 18 give some information about series which converge conditionally. Conditionally convergent series have almost paradoxical properties as we will see in the discussion which follows.

Definition 10: *Let $\sum a_i$ be any series. If $k: \mathbb{N} \to \mathbb{N}$ is both one-one and onto, then denoting $k(i)$ by k_i, the series*

$$\sum a_{k_i}$$

*is said to be a **rearrangement** of $\sum a_i$.*

A rearrangement of a series then is just a series formed by reordering the sequence on which the original series is based.

Proposition 20: If $\sum a_i$ is a conditionally convergent series and L is any real number, then there is a rearrangement of $\sum a_i$ which converges to L as well as a rearrangement of $\sum a_i$ which diverges.

PROOF: Let $p_1, p_2, \ldots$ be the positive terms and $q_1, q_2, \ldots$ the non-positive terms of $\sum a_i$ arranged in the order in which they occur as a_i. Now both $\sum p_i$ and $\sum q_i$ must diverge. For if $\sum p_i \to M$ and $\sum q_i \to M'$, then $\sum |a_i|$ is a monotone increasing sequence which is bounded above by $M + (-M')$; hence $\sum |a_i|$ converges, contradicting the assumption that $\sum a_i$ does not converge absolutely. And if either $\sum p_i$ or $\sum q_i$ diverges, while the other converges, then $\sum a_i$ diverges (the proof is left as an exercise), again a contradiction.

Let r_1 be the first positive integer for which $\sum_{i=1}^{r_1} p_i > L$. Let s_1 be the first positive integer for which $\sum_{i=1}^{r_1} p_i + \sum_{i=1}^{s_1} q_i < L$. Let r_2 be the first positive integer greater than r_1 such that

$$\sum_{i=1}^{r_1} p_i + \sum_{i=1}^{s_1} q_i + \sum_{i=r_1+1}^{r_2} p_i > L. \tag{15}$$

Let s_2 be the first positive integer greater than s_1 for which the left side of (15) plus $\sum_{i=s_1+1}^{s_2} q_i$ is less than L. Continue to define s_n and r_n in like fashion. We thus obtain a rearrangement

$$p_1, \ldots, p_{r_1}, q_1, \ldots, q_{s_1}, p_{r_1+1}, \ldots, p_{r_2}, q_{s_1+1}, \ldots, q_{s_2}, \ldots$$

of the terms of the sequence on which $\sum a_i$ is based; and hence we have a rearrangement of $\sum a_i$. Since $\sum a_i$ is a Cauchy sequence (since $\sum a_i$ converges), it follows that the rearrangement converges to L. The details of the proof of this latter statement are left as an exercise.

We also leave the construction of a divergent rearrangement of $\sum a_i$ as an exercise.

Proposition 21: A series $\sum a_i$ converges absolutely if and only if all rearrangements of $\sum a_i$ converge to the same limit. (This proposition is often taken as the definition of absolute convergence.)

PROOF: We first suppose that $\sum a_i$ converges absolutely. Let $\sum a_{k_i}$ be any rearrangement of $\sum a_i$. We denote the nth partial sums of $\sum a_i$ and $\sum a_{k_i}$ by t_n and t_{k_n}, respectively. Since $\sum a_i$ converges absolutely, given any $p > 0$, there is an integer M such that if $M < m \leqslant n$, then $\sum\limits_{i=m}^{n} |a_i| < p$. Choose an integer M' large enough so that $1, \ldots, M$ are all elements of $\{k_1, k_2, \ldots, k_{M'}\}$. Then if $n > M'$, $t_n - t_{k_n}$ will be a sum of terms which include $a_1 - a_1, a_2 - a_2, \ldots, a_M - a_M$ (since $a_1, \ldots, a_M$ will be terms of both t_n and t_{k_n}). This in turn implies that $|t_n - t_{k_n}| < p$, from which it follows that t_n and t_{k_n} both converge to the same limit.

If $\sum a_i$ does not converge absolutely, then rearrangements can be found which converge to any real number. This completes the proof.

We conclude this chapter with a discussion of multiplication of series.

Consider the series $\sum a_i$ and $\sum b_i$. If we think of these series as "infinite sums," $a_1 + a_2 + a_3 + \ldots$ and $b_1 + b_2 + b_3 + \ldots$, and apply the usual multiplication rules to these strings of symbols, then the product of $\sum a_i$ and $\sum b_i$ should contain all possible terms of the form $a_i b_j$, where i and j are positive integers. The real question is one of how these terms are to be arranged to form the product since different arrangements of the terms may lead to different sums. If some arrangement of the $a_i b_j$ converges absolutely, then all arrangements will converge to the same limit (Proposition 21). We will choose one classical arrangement which usually serves as the definition of the product of $\sum a_i$ and $\sum b_i$ and then prove one of its basic properties.

Definition 11: *Given the series $\sum a_i$ and $\sum b_i$, we define the **product** of $\sum a_i$ and $\sum b_i$, denoted by $\sum a_i \sum b_i$, to be $\sum c_i$, where*

$$c_i = \sum_{j=1}^{i} a_j b_{(i+1)-j}. \tag{16}$$

Greater justification for Definition 11 will be available after we consider power series.

Proposition 22: Suppose $\sum a_i$ converges absolutely to A and $\sum b_i$ converges to B. Then $\sum a_i \sum b_i$ converges to AB.

PROOF: We let u_n, s_n, and t_n represent the nth partial sums of $\sum a_i$, $\sum b_i$, and $\sum c_i$, respectively, with c_i as defined in (16): set $B_n = s_n - B$.

Then

$$t_n = a_1b_1 + (a_1b_2 + a_2b_1) + \ldots + (a_1b_n + a_2b_{n-1} + \ldots + a_nb_1)$$
$$= a_1s_n + a_2s_{n-1} + \ldots + a_ns_1$$
$$= a_1(B + B_n) + a_2(B + B_{n-1}) + \ldots + a_n(B + B_1)$$
$$= u_nB + a_1B_n + a_2B_{n-1} + \ldots + a_nB_1.$$

Now set

$$R_n = a_1B_n + a_2B_{n-1} + \ldots + a_nB_1. \tag{17}$$

Since we want to show that $t_n \to AB$ and since $u_nB \to AB$, it suffices to prove that $R_n \to 0$.

Since $\sum a_i$ converges absolutely, let L be the limit of $\sum |a_i|$. Suppose $p > 0$. We already know that $B_n \to 0$ since $s_n \to B$. Therefore there is an integer M such that $n \geqslant M$ implies $|B_n| < p$. Consequently,

$$|R_n| \leqslant |B_1a_n + \ldots + B_Ma_{n-M}| + |B_{M+1}a_{n-M-1} + \ldots + B_na_1|$$
$$\leqslant |B_1a_n + \ldots + B_Ma_{n-M}| + pL.$$

Since $a_n \to 0$ and p was arbitrary, we see that R_n must converge to 0.

EXERCISES

1. The following refer to the proof of Proposition 20.

a) Prove that if one of the series $\sum p_i$ and $\sum q_i$ converges while the other diverges, then $\sum a_i$ diverges.

b) Prove that the rearrangement constructed does in fact converge to L.

c) Construct a rearrangement of $\sum a_i$ which diverges.

2. Determine for which values of x the following series converge absolutely.

a) $\sum x^i$ d) $\sum x^i/2^i$

b) $\sum ix^i$ e) $\sum x/i$

c) $\sum x^i/i!$ f) $\sum \sqrt{x}/\sqrt{i}$

3. Prove or disprove: An alternating series $\sum a_i$ converges if and only if $a_i \to 0$. (Note that we have omitted the condition that $\{|a_n|\}, n \in \mathbb{N}$, is a monotone decreasing sequence.)

4. Find the first four terms of $(\sum 1/n)^2$.

5. Prove that if $\sum a_i$ converges and $\{b_n\}$ is a monotone bounded sequence, then $\sum a_i b_i$ converges.

6. Prove that the product of two absolutely convergent series converges absolutely.

7. If $\sum a_n$ is only conditionally convergent, is it possible for $\sum a_n^2$ to converge? Will $\sum a_n^2$ always converge?

8. Prove that the sum of $\sum\limits_{n=1}^{\infty} (-1)^{n+1}(1/n)$ lies between $1/2$ and 1.

5

THE DERIVATIVE

5.1 THE LIMIT OF A FUNCTION

LIMITS

We have already talked about the limit of a sequence; the limit of a sequence is that number which the terms of the sequence "approximate" as n gets larger and larger. A function $f: W \to \mathbb{R}$ may also have values which "approximate" some "limit" as x, the function argument, approaches some number a. For example, we expect $f(x) = x^2$ to "approach" 4 as x "approaches" 2.

We make the notion of a limit of a function more precise in the following definition.

Definition 1: *We say that the function $f: W \to \mathbb{R}$ has **limit** L **as** x **approaches** a if a is a limit point of W and given any $p > 0$, there is $q > 0$ such that $0 < |x - a| < q$ and $x \in W$ imply $|f(x) - L| < p$. If f has limit L as x approaches a we write $\lim_{x \to a} f(x) = L$.* *

Definition 1 says that $\lim_{x \to a} f(x) = L$ if given any $p > 0$, then if $x \in W$ sufficiently close to a, but not equal to a, then $f(x)$ will be within distance p of L. Because a is a limit point of W, there are points of the domain of f arbitrarily close to a, yet distinct from a. We do not want to allow the possibility of $x = a$ since this would virtually destroy an effective notion of limit; for example, consider the following.

---

* If the condition that a is a limit point of w is omitted, then if a is not a limit point of w, we would have $\lim_{x \to a} f(x) = L$ for *every* real number L.

Example 1:   Define $f(x) = 0$ for $x \neq 0$ and $f(0) = 1$. It is intuitively clear from the definition and graph of f that $\lim_{x \to 0} f(x)$ should be 0. Yet if we allowed the possibility of $x = 0$ in Definition 1, there would always be a number x, namely, $x = 0$, q-close to 0 for any $q > 0$, such that $|f(x) - 0| =$

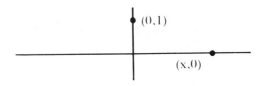

Figure 15

$|f(0) - 0| = 1 > p$ for any p less than 1. We leave it to the reader to prove that $\lim_{x \to 0} f(x) = 0$ according to Definition 1.

The concepts of limit of a function and continuity are closely related. In particular, we have the following proposition.

Proposition 1:   If $f: W \to \mathbb{R}$, then $\lim_{x \to a} f(x) = L$ if and only if given any sequence $\{s_n\}$, $n \in \mathbb{N}$, in $W \sim \{a\}$ such that $s_n \to a$, then $f(s_n) \to L$.

PROOF: Once again the proof is similar to that of Proposition 19 of Chapter 3. First suppose that $\lim_{x \to a} f(x) = L$ and $\{s_n\}$, $n \in \mathbb{N}$, is a sequence of points of $W \sim \{a\}$ which converges to a. Let $p > 0$; then there is $q > 0$ such that $0 < |x - a| < q$ and $x \in W$ implies $|f(x) - L| < p$. Since $s_n \to a$, all but finitely many s_n are in $N(a, q) \cap W$. Therefore all but finitely many $f(s_n)$ are in $N(L, p)$. Consequently, $f(s_n) \to L$.

If $\lim_{x \to a} f(x) \neq L$, then a sequence $\{s_n\}$, $n \in \mathbb{N}$, of points of $W \sim \{a\}$ can be constructed which converges to a, but such that $f(s_n) \nrightarrow L$. We leave it to the reader to construct such a sequence.

Corollary 1:   If $f: W \to \mathbb{R}$, then f is continuous at $a \in W$ if and only if $\lim_{x \to a} f(x) = f(a)$, or a is not a limit point of W.
We leave the proof of this corollary as an exercise.

Because limits of sequences are unique, we also have the following corollary.

Corollary 2:   If $\lim_{x \to a} f(x) = L$ and $\lim_{x \to a} f(x) = L'$, then $L = L'$; that is, a function can have at most one limit as x approaches a.

Proposition 2: If f and g are functions from W into $\mathbb{R}$ such that $\lim\limits_{x \to a} f(x) = L$ and $\lim\limits_{x \to a} g(x) = L'$, then

a) $\lim\limits_{x \to a} (f + g)(x) = L + L'$,

b) $\lim\limits_{x \to a} (fg)(x) = LL'$,

c) $\lim\limits_{x \to a} (rf)(x) = rL$,

d) $\lim\limits_{x \to a} (1/f)(x) = 1/L$ provided $L \neq 0$.

Proposition 2 follows almost immediately from Proposition 1 of this chapter and Proposition 2 of Chapter 4. We leave the actual proof to the reader.

We now give some example of functions which have limits and some which do not have limits.

Example 2: Consider $f(x) = x^3$ defined on $(0, 1)$. Then it can be proved formally using Definition 1 that $\lim\limits_{x \to 1} f(x) = 1$. We can also note that while f is technically defined only on the open interval $(0, 1)$ (because that is the way we defined it), it can be extended using the same definition to a function which is defined and continuous on all of $\mathbb{R}$. This being the case, we will have $\lim\limits_{x \to 1} f(x) = f(1) = 1$, where f is now considered on all of $\mathbb{R}$.

Example 3: The function $f(x) = 1/x$ does not approach any limit as $x \to 0$ since it does not remain bounded as $x \to 0$ (see Exercise 10). There is no way to define $f(0)$ so as to extend f to be a continuous function for all of $\mathbb{R}$. If $\lim\limits_{x \to 0} f(x)$ did exist and was, say, L, then we could define $f(0) = L$ to obtain a continuous definition of f for all of $\mathbb{R}$.

Example 4: Consider $f(x) = \sin(1/x)$ defined for $x \neq 0$. (We will discuss the trigonometric functions formally later in this text. We assume, however, that the reader has some familiarity with the sine and cosine and their graphs as well as radian angle measurement.) The function f is bounded since the sine of any number lies between -1 and 1. The graph of f is given in Figure 16. Even though f is bounded, f does not approach a limit as $x \to 0$. For consider the sequence defined

$$s_n(y) = 1/(2\pi n + y), \tag{1}$$

where y is an arbitrary positive number. The sequence defined in (1) converges to 0, and the sequence of $f(s_n(y)) = \sin(2\pi n + y) = \sin y$ converges to $\sin y$. Now y is an arbitrary positive number; hence there are distinct

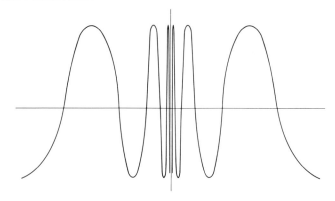

Figure 16

sequences $\{s_n(y)\}$, $n \in \mathbb{N}$, and $\{s_n(y')\}$, $n \in \mathbb{N}$, with $\sin y \neq \sin y'$, both of which converge to 0. Since $f(s_n(y)) \to \sin y$ and $f(s_n(y')) \to \sin y'$, but $\sin y \neq \sin y'$, $\lim_{x \to 0} f(x)$ does not exist by Proposition 1 and Corollary 2 to Proposition 1.

Example 5: Consider

$$\lim_{h \to 0} \frac{(1/(3 + h) - 1/3)}{h} . \tag{2}$$

Trying to substitute $h = 0$ in (2) gets us nowhere since we then obtain the meaningless expression $0/0$. The fractional part of (2) is equivalent to

$$\frac{(3 - (3 + h))/(3(3 + h))}{h} = -(h/h)(1/[3(3 + h)]) = -1/(3(3 + h)). \tag{3}$$

The limit of the last expression in (3) as $h \to 0$ is clearly $-1/9$. Therefore (2) is $-1/9$. Although (2) could not be evaluated exactly as given, suitable algebraic manipulations enabled us to evaluate the limit.

THE FIRST DERIVATIVE

Expression (2) is a special case of an important class of limits, the *first derivatives*, which we will now investigate.

Definition 2: *Let* $f: W \to \mathbb{R}$ *and* $a \in W$. *We assume that some open interval which contains a is a subset of* W. *Then if*

$$\lim_{h \to 0} \frac{f(a + h) - f(a)}{h} \tag{4}$$

*exists, we call this limit the **first derivative** of f at a, and we say that f is **differentiable at a.** We denote the first derivative of f at a by f′(a).*

*If f is differentiable at each a ∈ U ⊆ W, then f is said to be **differentiable on** U. If U = W, then f is said to be **differentiable.***

*If a is an end point of W (for example, if W is the closed interval [a, b]), the f is said to be **differentiable** at a if f can be extended to an open interval which contains a such that the extended function is differentiable at a. (That is, f is differentiable at a if there is a function g defined on some open interval U which contains a such that g′(a) exists and g(x) = f(x) for all x ∈ U ∩ W. Cf. Exercise 6.)*

Geometrically speaking, $(f(a + h) - f(a))/h$ is the slope of the secant line to the graph of f passing through the points $(a, f(a))$ and $(a + h, f(a + h))$ (Fig. 17). Therefore $f'(a)$ is the limiting slope of such secant lines as $h \to 0$,

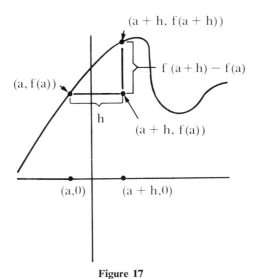

Figure 17

that is, as $a + h \to a$. We can therefore consider (or define) $f'(a)$ to be the slope of the line *tangent* to the graph of f at $(a, f(a))$.

Example 6: The limit found in Example 5 is $f'(3)$ for $f(x) - 1/x$. Compare (2) with (4) letting $f(x) = 1/x$.

Differentiable functions have particularly "nice" properties; they are even more "well-behaved" than continuous functions. A function may be continuous and fail to be differentiable, but no function can be differentiable and fail to be continuous (cf. Example 7 and Proposition 3). Therefore for any subset W of $\mathbb{R}$, the functions which are differentiable on W form a subset of the continuous functions defined on W.

Example 7:   Consider the function $f: \mathbb{R} \to \mathbb{R}$, where $f(x) = |x|$. The graph of f is given in Figure 18. Note that although this function is con-

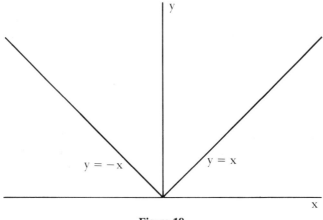

Figure 18

tinuous at 0, the graph of f at 0 is not "smooth;" there is a sharp corner at $(0, 0)$. We now show that f does not have a first derivative at 0. Consider

$$(f(0 + h) - f(0))/h = |h|/h. \tag{5}$$

Now if $h > 0$, then (5) is 1, while if $h < 0$, then (5) is -1. Therefore the limit of (5) as $h \to 0$ depends on how h approaches 0; that is, if h approaches 0 from the negative side, then we would expect $\lim_{h \to 0} (|h|/h) = -1$, while from the positive side this limit is 1. Therefore the limit of (5) as $h \to 0$ does not exist (if it existed, by the uniqueness of the limit the limit should be the same regardless of how h approaches 0), therefore $f'(0)$ does not exist.

Proposition 3:   If $f'(a)$ exists, then f is continuous at a.

PROOF: Consider (4). If f is not continuous at a, then $\lim_{h \to 0} f(a + h)$ does not equal $f(a)$ (for if it did, then we would have $\lim_{x \to a} f(x) = f(a)$ and then f would be continuous at a). Therefore the numerator of (4) does not go to 0 as h goes to 0; hence the limit in (4) could not exist.

There are functions which are defined and continuous for each real number, but which are not differentiable at any real number. Although such functions are infinite in number, the proof of the existence of such a function is beyond the pretensions of this text. Related to each differentiable function, there is a function called its first derivative.

Definition 3: *If $f: W \to \mathbb{R}$ is differentiable, then the function f', where $f'(a)$ is the first derivative of f at a for each $a \in W$, is called the **first derivative** of f.*

Although f' is defined on all of W if $f: W \to \mathbb{R}$ is differentiable, it is quite possible for f' to be non-differentiable; in fact, f' may not even be continuous. Specific examples of such behavior by f' must wait until later in the text.

EXERCISES

1. In Example 1, prove that $\lim_{x \to 0} f(x) = 0$.

2. Prove Corollary 1 of Proposition 1.

3. Construct the sequence called for in the second part of the proof of Proposition 1.

4. Prove any two parts of Proposition 2.

5. Find the first derivative called for at the point indicated in each of the following. Compute the derivative directly using Definition 3.

a) $f(x) = x$; $x = 3$ c) $f(x) = x^{1/2}$; $x = 1$
b) $f(x) = x^2$; $x = 3$ d) $f(x) = 4x - 2$; $x = -5$

6. Let $f: W \to \mathbb{R}$ and suppose $a \in W$ such that there is $b \in W$ with $[a, b) \subseteq W$. Define

$$f'_+(a) = \lim_{\substack{h \to 0 \\ h > 0}} (f(a + h) - f(a))/h$$

(that is, $a + h$ approaches a from inside $[a, b)$.

a) How would we define $f'_-(a)$ if there were $b \in W$ such that $(b, a] \subseteq W$?
b) Compute $f'_+(0)$ and $f'_-(0)$, where $f(x) = |x|$.
c) Prove or disprove: If a is contained in an open interval contained in W, then $f'(a)$ exists if and only if $f'_+(a)$ and $f'_-(a)$ exist and are equal.
d) Prove that if a is the left end point of W, then $f'_+(a)$ exists if and only if f is differentiable at a.

7. Find a function which is continuous on $(0, 1)$, but which is not differentiable at $x = 1/n$, n any positive integer.

8. Find a function defined on all of $\mathbb{R}$, but which is continuous and differentiable only at 0. Hint: Let S and T be the sets of rational and irrational numbers, respectively; define $f(x) = x^2$ for $x \in S$ and $f(x) = 0$ for $x \in T$.

9. We say $\lim_{x \to \infty} f(x) = L$ if given any $p > 0$, there is a positive number M such that $x > M$ implies $|f(x) - L| < p$. Define what is meant by $\lim_{x \to -\infty} f(x) = L$. We say that $\lim_{x \to a} f(x) = \infty$ if given any positive number M there is $q > 0$ such that $|x - a| < q$ implies $f(x) > M$. Define what is meant by $\lim_{x \to a} f(x) = -\infty$. Find each of the following limits.

a) $\lim_{x \to 0} 1/x^2$

b) $\lim_{x \to \infty} 1/x^2$

c) $\lim_{x \to -\infty} 1/x$

d) $\lim_{x \to \infty} x^{1/2}/(x + 1)$

e) $\lim_{x \to -1} 1/(1 - x^2)$

10. Prove that if $\lim_{x \to a} f(x) = L$, then f must be bounded on some open set which contains a.

5.2 MORE BASIC FACTS ABOUT THE FIRST DERIVATIVE

In general, it is at best a tedious job to find the first derivative of a function directly from Definition 2. The definition might be directly applicable to such simple functions as $f(x) = x^3$, but difficulties become apparent when trying to compute the derivative of $g(x) = (x^{24} + 17x^{14} - 9x + 1)^{798}$. We begin this section by proving some facts which greatly simplify the taking of many derivatives.

Proposition 4: If $f: W \to \mathbb{R}$ is defined by $f(x) = k$ for all $x \in W$, where k is a constant, then $f'(a) = 0$ for any a contained in an open interval which is a subset of W.

We leave the proof of Proposition 4 to the reader.

Proposition 5: Suppose f and g are both functions from W into $\mathbb{R}$ which are differentiable at a. Then $(f + g)'(a) = f'(a) + g'(a)$. (That is, the first derivative of the sum of two functions is the sum of the first derivatives.)

PROOF

$$(f + g)'(a) = \lim_{h \to 0} \frac{(f + g)(a + h) - (f + g)(a)}{h}$$

$$= \lim_{h \to 0} \frac{(f(a + h) - f(a)) + (g(a + h) - g(a))}{h}$$

$$= \lim_{h \to 0} \frac{f(a + h) - f(a)}{h} + \lim_{h \to 0} \frac{g(a + h) - g(a)}{h}$$

$$\text{(cf. Proposition 2)}$$

$$= f'(a) + g'(a).$$

Proposition 6: If $f: W \to \mathbb{R}$ is differentiable at a and r is any real number, then $(rf)'(a) = rf'(a)$.

We leave the proof of Proposition 6 to the reader.

Proposition 7: If f and g are both functions which are differentiable at a, then $(fg)'(a) = f(a)g'(a) + g'(a)f(a)$. (Note that the derivative of a product is *not* the product of the derivatives.)

PROOF

$$\frac{(fg)(a + h) - (fg)(a)}{h} = \frac{f(a + h)g(a + h) - f(a)g(a)}{h}$$

$$= f(a + h) \frac{g(a + h) - g(a)}{h} \qquad (6)$$

$$+ g(a + h) \frac{f(a + h) - f(a)}{h}.$$

(Straightforward computation, if nothing else, confirms the last equality.) Since f and g are differentiable at a, they are continuous at a; hence $\lim_{h \to 0} f(a + h) = f(a)$ and $\lim_{h \to 0} g(a + h) = g(a)$. Applying the appropriate parts of Proposition 2, we have that the limit of (6) as $h \to 0$ is $f(a)g'(a) + g(a)f'(a)$, which is what we wanted to show.

Proposition 8: If n is any positive integer and $f: \mathbb{R} \to \mathbb{R}$ is defined by $f(x) = x^n$, then

$$f'(x) = nx^{n-1}, \qquad (7)$$

for any x in $\mathbb{R}$.

PROOF: Proposition 8 is readily verified for $n = 1$. Assume that Proposition 8 is true for n: we will show that this implies the proposition must be true for $n + 1$. Suppose $f(x) = x^{n+1}$; then $f(x) = g(x)x$, where

$g(x) = x^n$. By Proposition 7 and the induction assumption,

$$f'(x) = g(x)(1) + g'(x)x = x^n + nx^{n-1}x = (n + 1)x^n,$$

which is (7) applied to $f(x) = x^{n+1}$. By the First Principle of Finite Induction (Proposition 2(c) of Chapter 2), the proposition is true for each positive integer n.

Proposition 9 (*The Chain Rule*):   If f and g are functions such that g is differentiable at $f(x)$ and f is differentiable at x, then $(g \circ f)'(x) = g'(f(x))f'(x)$.

Before proving Proposition 11, we prove a simple lemma.

Proposition 10

$$\lim_{h \to 0} \frac{f(x + h) - f(x)}{h} = \lim_{t \to x} \frac{f(t) - f(x)}{t - x}.$$

PROOF: Proposition 1 follows at once by setting $t = x + h$.

We now prove Proposition 9.

PROOF OF PROPOSITION 9:   Since $f'(x) = \lim_{t \to x} (f(t) - f(x))/(t - x)$, we have $f(t) - f(x) = (t - x)(f'(x) + r(t))$, where $\lim_{t \to x} r(t) = 0$.   Similarly,

$$\frac{g(f(t)) - g(f(x))}{f(t) - f(x)} = g'(f(x)) + s(f(t)),$$

where $\lim_{f(t) \to f(x)} s(f(t)) = 0$.   But since $f(t) \to f(x)$ as $t \to x$, we also have $\lim_{t \to x} s(f(t)) = 0$.   We therefore have

$$g(f(t)) - g(f(x)) = (f(t) - f(x)) \times (g'(f(x)) + s(f(t)))$$
$$= (t - x)(f'(x) + r(t))(g'(f(x)) + s(f(t))),$$

from which we obtain

$$\frac{g(f(t)) - g(f(x))}{t - x} = (g'(f(x)) + s(f(t)))(f'(x) + r(t)). \tag{8}$$

Taking the limit of both sides of (8) as $t \to x$, we obtain $(g \circ f)'(x) = g'(f(x))f'(x)$, which is what we wished to prove.

Proposition 11:   If f and g are both differentiable functions at x and if $g(x) \neq 0$, then

$$(f/g)'(x) = \frac{g(x)f'(x) - g'(x)f(x)}{(g(x))^2}.$$

PROOF

$$\frac{(f/g)(x+h) - (f/g)(x)}{h} = \frac{f(x+h)/g(x+h) - f(x)/g(x)}{h}$$

$$= \frac{1}{g(x+h)g(x)}\left(g(x)\frac{f(x+h) - f(x)}{h}\right. \tag{9}$$

$$\left. - f(x)\frac{g(x+h) - g(x)}{h}\right).$$

Taking the limit of (9) as $h \to 0$, we obtain the desired result.

Proposition 12: If n is any positive integer and if $f: \mathbb{R} \sim \{0\} \to \mathbb{R}$ is defined by $f(x) = x^{-n}$, then $f'(x) = (-n)x^{-n-1}$ for all x in $\mathbb{R} \sim \{0\}$.

PROOF: If $n = -1$, then $f(x) = 1/x$. Considering f as g/h, where $g(x) = 1$ and $h(x) = x$, and applying Proposition 11, we find $f'(x) = -1/x^2 = (-1)(x^{-2})$, which proves Proposition 12 for $n = 1$. Now if $f(x) = x^{-n}$, $n \geqslant 2$, we have $f(x) = (g(x))^n$, where $g(x) = x^{-1}$. Using the derivative of g which we have just found and the Chain Rule, we see that $f'(x) = n(g(x))^{n-1}g'(x) = n(x^{-1})^{n-1}(-1)(x^{-2}) = (-n)x^{-n-1}$, which is what we wanted to prove.

Corollary: For any integer n, if $f(x) = x^n$, $f'(x) = nx^{n-1}$.

Proposition 13: Suppose f is differentiable (and hence defined and continuous) and strictly increasing (or strictly decreasing) on an open interval U, $a \in U$, and $f(a) = b$. As we saw in Proposition 37 of Chapter 4, $f^{-1}:f(U) \to \mathbb{R}$ is a continuous function. If $f'(a) \neq 0$, then f^{-1} is differentiable at b and $(f^{-1})'(b) = 1/f'(a)$.

PROOF: We want to prove

$$\lim_{y \to b} \frac{f^{-1}(y) - f^{-1}(b)}{y - b} = 1/f'(a) . \tag{10}$$

Set $x = f^{-1}(y)$. Then $y = f(x)$; therefore

$$\frac{f^{-1}(y) - f^{-1}(b)}{y - b} = \frac{x - a}{f(x) - f(a)} = \frac{1}{\dfrac{f(x) - f(a)}{x - a}} .$$

Since

$$1/f'(a) = \frac{1}{\displaystyle\lim_{x \to a} \frac{f(x) - f(a)}{x - a}} = \lim_{x \to a} \frac{1}{\dfrac{f(x) - f(a)}{x - a}},$$

(10) will follow provided $x \to a$ as $y \to b$, that is $f(x) \to f(a)$ as $x \to a$, or $\lim_{x \to a} f(x) \to f(a)$. Since this latter statement follows from the continuity of f, Proposition 13 is proved.

We use Proposition 13 in the following example.

Example 8: The function f defined by $f(x) = x^n$, n a positive integer, is strictly increasing for $x > 0$; moreover, for $x > 0$, $f'(x) \neq 0$. Therefore f^{-1} defined by $f^{-1}(y) = y^{1/n}$ is differentiable for $y \in f(\{x \mid x > 0\})$, that is, for $y > 0$. By Proposition 13, $(f^{-1})'(y) = 1/f'(x) = 1/(nx^{n-1})$, where $f(x) = y$. Since $y = x^n$, $x = y^{1/n}$; hence $1/f'(x) = (1/n)(1/(y^{1/n})^{n-1} = (1/n)y^{-(n-1)/n}$. We therefore conclude that if $y > 0$ and g is the function defined by $g(y) = y^{1/n}$, then $g'(y) = (1/n)y^{-(n-1)/n}$.

The following proposition will not be proved until after the exponential function and logarithms have been discussed in Chapter 7. We include this proposition now for the sake of completeness.

Proposition 14: If r is any real number and if f is the function defined by $f(x) = x^r$ for $x > 0$, then $f'(x) = rx^{r-1}$.

As yet we do not even have a definition for x^r when r is irrational (although we are in a position to work with x^r if r is rational; see Exercise 4). This definition will come in Chapter 7.

EXERCISES

1. Find the first derivative of each of the following functions.

a) $f(x) = x + x^3$
b) $f(x) = (x^{1/2} + 1)^{1/2}$
c) $f(x) = (x + 1)^{\pi}x^{-6}$
d) $f(x) = (((x + 1)^2 + 1)^2 + 1)^2$
e) $f(x) = (x - 3)/(x^2 - 4)$
f) $f(x) = ((x^2 - x^{1/2})/(x^5 - 1)^3)^{1/3}$

2. Prove Proposition 4.

3. Prove Proposition 6.

4. Prove that if $r = p/q$, where p and q are non-zero integers, and if f is defined by $f(x) = x^r$, $x > 0$, then $f'(x) = rx^{r-1}$ for all $x > 0$.

5. Suppose $f: \mathbb{R} \sim \{0\} \to \mathbb{R}$ is defined by $f(x) = 1/x$. Prove directly from Definition 2 that f is differentiable for all x for which it is defined and that $f'(x) = -1/x^2$. Use this result and the Chain Rule to prove Proposition 11.

5.3 SOME USES OF THE FIRST DERIVATIVE

Since the reader is assumed to have already had some experience with calculus, he should recognize some of the following propositions as being basic to certain practical applications of the calculus. We are not so much interested in the applications now as in proving the statements rigorously.

Definition 4: *A function f is said to be **increasing** on an interval I if f is defined at each point of I and for each x and y in I with $x \leqslant y$, $f(x) \leqslant f(y)$. (The definition for **decreasing** at a is similar to the preceding except that $x \leqslant y$ implies $f(y) \leqslant f(x)$.)*
*The function f is said to have a **relative maximum** at a if there is $p > 0$ such that f is defined in $(a - p, a + p)$ and $f(x) \leqslant f(a)$ for all x in $(a - p, a + p)$. We leave it to the reader to formulate the definition of a **relative minimum** of f.*

Proposition 15: If f has a relative maximum at a and is differentiable at a, then $f'(a) = 0$.

PROOF: We can find $p > 0$ such that f is defined on $(a - p, a + p)$ and $f(x) \leqslant f(a)$ for each $x \in (a - p, a + p)$. Consider

$$(f(t) - f(a))/(t - a). \tag{11}$$

If $a - p < t < a$, then $t - a$ is negative and $f(t) - f(a)$ is non-positive; hence (11) is non-negative. On the other hand, if $a < t < a + p$, then $t - a$ is positive, but $f(t) - f(a)$ is still non-positive; hence (11) is non-positive. Since the limit of (11) as $t \to a$ is $f'(a)$, $f'(a)$ must be both non-negative and non-positive It follows then that $f'(a)$ is 0.

The proof of the following proposition is quite similar to the preceding proof. We leave this proof to the reader.

Proposition 16: If f has a relative minimum and is differentiable at a, then $f'(a) = 0$.

We now prove the *First Mean Value Theorem.*

Proposition 17:   Suppose f is a function which is continuous on $[a, b]$ and f is differentiable on (a, b). Then there is a point x' of (a, b) such that

$$\frac{f(b) - f(a)}{b - a} = f'(x').$$

(Interpreted geometrically, this proposition states that there is a line tangent to the graph of f at some point in $[a, b]$ which is parallel to—has the same slope as—the line determined by the end points of the graph over $[a, b]$; cf. Figure 19.)

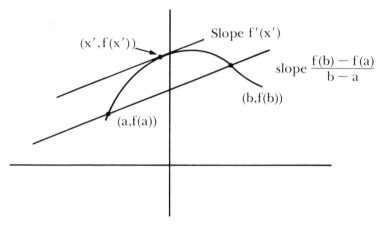

Figure 19

PROOF: We define the function F on $[a, b]$ by setting

$$F(x) = f(x) - f(a) - \frac{f(b) - f(a)}{b - a}(x - a).$$

Then F is continuous on $[a, b]$ and differentiable on (a, b) because of the corresponding properties of f. Specifically,

$$F'(x) = f'(x) - \frac{f(b) - f(a)}{b - a}.$$

Thus we would like to show that there is x' in (a, b) such that $F'(x') = 0$. We note that $F(a) = F(b) = 0$. If $F(x) \neq 0$ for some point of (a, b), then F has either a relative maximum or minimum at some point x' of (a, b) (Corollary to Proposition 29 of Chapter 3). Therefore $F'(x') = 0$ by Proposition 15 or 16. If $F(x) = 0$ for every x in (a, b), then $F'(x') = 0$ for any $x' \in (a, b)$.

Proposition 18: Suppose that f is differentiable on (a, b). Then

a) $f'(x) \geqslant 0$ for all $x \in (a, b)$ implies f is increasing on (a, b).

b) $f'(x) \leqslant 0$ for all $x \in (a, b)$ implies f is decreasing on (a, b).

c) $f'(x) = 0$ for all $x \in (a, b)$ implies f is constant on (a, b).

PROOF: If x and y are any two points of (a, b) with $x < y$, then by Proposition 17 there is x' with $x < x' < y$ such that

$$f(y) - f(x) = (y - x)f'(x'). \tag{12}$$

Since $y - x$ is positive, the sign of $f(y) - f(x)$ is determined by the sign of $f'(x')$, and the conclusions of the proposition follow at once.

Corollary: If f and g are differentiable on (a, b) and $f'(x) = g'(x)$ for all $x \in (a, b)$, then $f(x) = g(x) + k$, k a constant, for all $x \in (a, b)$.

Proposition 19 (*The Second Mean Value Theorem*): Suppose f and g are both continuous on $[a, b]$ and differentiable on (a, b), and suppose too that $g'(x) \neq 0$ for any $x \in (a, b)$. Then there is a point x' of (a, b) such that

$$\frac{f(b) - f(a)}{g(b) - g(a)} = \frac{f'(x')}{g'(x')} .$$

(Note that Proposition 19 generalizes Proposition 17; for if we set $g(x) = x$ then the statement of Proposition 19 is that of Proposition 17.)

PROOF: By Proposition 17 we can find $x'' \in (a, b)$ such that

$$g(b) - g(a) = (b - a)g'(x'').$$

Since $g'(x'') \neq 0$, it follows that $g(b) \neq g(a)$. We now define

$$F(x) = f(x) - f(a) - \frac{f(b) - f(a)}{g(b) - g(a)} (g(x) - g(a)) \text{ for all } x \in [a, b].$$

The function F satisfies the hypotheses of Proposition 17; moreover, $F(b) = F(a) = 0$. Therefore there is $x' \in (a, b)$ such that $F'(x') = 0$. Since

$$F'(x) = f'(x) - g'(x)\frac{f(b) - f(a)}{g(b) - g(a)} ,$$

Proposition 19 follows at once.

We now prove *L'Hôpital's Rule*, a proposition useful in the evaluation of certain limits.

Proposition 20:   Suppose f and g are both differentiable in $(0, p]$ for some $p > 0$ but $g'(x) \neq 0$ for all $x \in (0, p]$.   Then if

$$\lim_{x \to 0} f(x) = \lim_{x \to 0} g(x) = 0 \quad (x \to 0 \text{ from inside } (0, p]), \tag{13}$$

and if

$$\lim_{x \to 0} f'(x)/g'(x) \quad \text{exists and is equal to } L, \tag{14}$$

then $\lim_{x \to 0} f(x)/g(x) = L$.

PROOF: If we define $f(0) = g(0) = 0$, then f and g will both be continuous on $[0, p]$.   Therefore for any $x \in (0, p]$, there is $y \in (0, x)$ such that

$$\frac{f(x) - f(0)}{g(x) - g(0)} = \frac{f'(y)}{g'(y)}. \tag{15}$$

We cannot have $g(x) = g(0) = 0$ since if we did, there would be a point w of $(0, x)$ for which $g'(w) = 0$, contrary to the hypothesis that $g'(x) \neq 0$ for all $x \in (0, p]$.   Since $g(0) = f(0) = 0$, from (15) we have

$$\frac{f(y)}{g(y)} = \frac{f'(x)}{g'(x)}. \tag{16}$$

Since y must approach 0 as $x \to 0$, it follows from (16) that $\lim_{x \to 0} f(x)/g(x) = \lim_{x \to 0} f'(x)/g'(x) = L$.

Corollary 1:   If f and g are differentiable in $[-p, 0)$ for some $p > 0$ but $g'(x) \neq 0$ for any $x \in [-p, 0)$, then if (13) and (14) are satisfied but where now x approaches 0 from within $[-p, 0)$, then $\lim_{x \to 0} f(x)/g(x)$ exists and is equal to L.

Corollary 2:   If f and g are differentiable in $[-p, p]$ for some $p > 0$ but $g'(x) \neq 0$ for any $x \in (-p, p)$, then if (13) and (14) (omitting $x > 0$) are satisfied, $\lim_{x \to a} f(x)/g(x)$ exists and is L.

Corollary 3:   If f and g are differentiable in $(a - p, a + p)$ for $p > 0$ but $g'(x) \neq 0$ for any $x \in (a - p, a + p)$ and if $\lim_{x \to a} f(x) = \lim_{x \to a} g(x) = 0$, and if $\lim_{x \to a} f'(x)/g'(x)$ exists and is L, then $\lim_{x \to a} f(x)/g(x)$ exists and is L.

The proofs of these corollaries are left as exercises.

We have seen that if f is a continuous function on the closed interval $[a, b]$ and if c is a number between $f(a)$ and $f(b)$, there is $x \in [a, b]$ such that

$f(x) = c$; this was the conclusion of the so-called Intermediate Value Theorem. While it is not true that derivatives are necessarily continuous, they do satisfy the Intermediate Value Theorem. Specifically, we have the following result.

Proposition 21: Suppose that f is differentiable on $[a, b]$ and $f'(a) < c < f'(b)$. Then there is a point x' of (a, b) for which $f'(x') = c$. (The proposition is also true, of course, if $f'(b) < c < f'(a)$, the proof being essentially the same.)

PROOF: Set $g(x) = f(x) - cx$ for all x in $[a, b]$. Then $g'(a) = f'(a) - c$ is negative, while $g'(b) = f'(b) - c$ is positive. We need to show that $g'(x') = 0$ for some $x' \in (a, b)$. Now g is continuous on $[a, b]$ since it is differentiable; hence there is a point x' of $[a, b]$ for which $g(x')$ is the minimum value of g on $[a, b]$. We now show that $x' \in (a, b)$. Now g could not attain its minimum at a since it is decreasing at a (since $g'(a) < 0$), nor could it have its minimum at b since it is increasing at b, and hence $f(y) < b$ for some $y < b$. Therefore $x' \in (a, b)$. But by Proposition 16, then $g'(x') = 0$, which is what we wanted to show.

EXERCISES

1. Prove Proposition 16.

2. Prove the corollaries to Proposition 19.

3. Prove that if (i) f is differentiable in some open interval (u, v) which contains a, (ii) f' is differentiable on (u, v), (iii) $(f')'$ or f'' is continuous on (u, v), and (iv) $f'(a) = 0$ while $f''(a) < 0$, then f has a relative minimum at a. Make and prove the corresponding statement for $f'(a) = 0$ and $f''(a) > 0$.

4. Let f be a function which is defined in $(a - p, a + p)$, $p > 0$, but which is discontinuous at a. We say that f has a *simple discontinuity* at a if $\lim_{\substack{x \to a \\ x > a}} f(x)$ and $\lim_{\substack{x \to a \\ x < a}} f(x)$ both exist but are not equal.

a) Prove that if f is the first derivative of a function g on $(a - p, a + p)$ then if f' fails to be continuous at a, the discontinuity cannot be simple.

b) Prove that a monotone increasing function can have only simple discontinuities.

5. Prove Proposition 20 with the following assumption replacing (13).

$$\lim_{\substack{x \to 0 \\ x > 0}} f(x) = \lim_{\substack{x \to 0 \\ x > 0}} g(x) = \infty. \tag{13'}$$

(Cf. Exercise 9 of Section 5.1.)

6. Prove that if f is differentiable on (a, b) and f' is positive on (a, b), then f is one-one on (a, b). State and prove the corresponding statement of f' negative on (a, b).

7. Prove the corollary to Proposition 18.

8. Provide examples to show that both hypotheses of the First Mean Value Theorem are necessary for its conclusion. For example, demonstrate a function which is differentiable on (a, b) but not continuous on $[a, b]$ for which there is no x' in (a, b) such that

$$f'(x') = \frac{f(b) - f(a)}{b - a}.$$

6

THE RIEMANN
INTEGRAL

6.1 THE DEFINITION OF THE RIEMANN INTEGRAL

Definition 1: *Let $[a, b]$ be any closed interval. By a* **partition** *of $[a, b]$ we mean a sequence of numbers $x_0, x_1, \ldots, x_n$ such that*

$$a = x_0 \leqslant x_1 \leqslant \ldots \leqslant x_n = b.$$

If $P = \{x_0, x_1, \ldots, x_n\}$ is a partition of $[a, b]$, we define the **mesh** *of P to be the maximum of the quantities $x_{i+1} - x_i$, $i = 0, 1, \ldots, n - 1$. We will denote the mesh of a partition P by $m(P)$.*

Geometrically, a partition of $[a, b]$ is a subdivision of $[a, b]$ into closed intervals which overlap only at end points. The mesh of a partition is the length of the largest of these subintervals.

We recall that a function on $[a, b]$ is said to be *bounded* if $f([a, b])$ is a bounded set. If $P = \{x_0, x_1, \ldots, x_n\}$ is a partition of $[a, b]$ and f is bounded on $[a, b]$, then f is also bounded on each subinterval $[x_{i+1}, x_i]$ of $[a, b]$. We set

$$M_i = \text{lub } \{f(x) \mid x \in [x_i, x_{i+1}]\},$$

and   (1)

$$m_i = \text{glb } \{f(x) \mid x \in [x_i, x_{i+1}]\}.$$

For the remainder of this chapter we will assume, unless specifically noted to the contrary, that f is a function which is defined and bounded on $[a, b]$.

Definition 2:   *For any partition $P = \{x_0, x_1, \ldots, x_n\}$ of $[a, b]$, set*

$$U(P, f) = \sum_{i=0}^{n-1} M_i(x_{i+1} - x_i), \tag{2}$$

and

$$L(P, f) = \sum_{i=0}^{n-1} m_i(x_{i+1} - x_i). \tag{3}$$

We call (2) *and* (3) *the* **upper** *and* **lower Riemann sums of f over $[a, b]$ relative to P,** *respectively.**

We set

$$\overline{\int}_a^b f(x)\,dx = \text{glb } \{U(P, f) \mid P \text{ a partition of } [a, b]\}, \tag{4}$$

and

$$\underline{\int}_a^b f(x)\,dx = \text{lub } \{L(P, f) \mid P \text{ a partition of } [a, b]\}. \tag{5}$$

We call (4) *and* (5) *the* **upper** *and* **lower Riemann integrals of f over $[a, b]$,** *respectively.*

If (4) *and* (5) *are equal, then we say that f is* **Riemann integrable,** *or simply* **integrable,** *over $[a, b]$. If f is integrable over $[a, b]$, then we call the common value of* (4) *and* (5) *the* **Riemann integral,** *or the* **definite integral of f over $[a, b]$.** *We denote the definite integral of f over $[a, b]$ by $\int_a^b f(x)\,dx$.*

Example 1:   Suppose f is a monotone increasing function on $[a, b]$, and $P = \{x_0, x_1, \ldots, x_n\}$ is a partition of $[a, b]$. Then since f is increasing on $[x_i, x_{i+1}]$, $i = 0, \ldots, n - 1$, we will always have $m_i = f(x_i)$ and $M_i = f(x_{i+1})$. Thus, in this instance,

$$U(P, f) = \sum_{i=0}^{n-1} f(x_{i+1})(x_{i+1} - x_i),$$

and

$$L(P, f) = \sum_{i=0}^{n-1} f(x_i)(x_{i+1} - x_i).$$

If P is a partition of $[a, b]$, $U(P, f)$ and $L(P, f)$ can be represented geometrically as in Figure 20. This figure should, however, be taken more as "indicative" than "definitive."

That (4) and (5) always exist is evident from the following proposition.

---

* Georg Bernhard Riemann (1826–1866) was a German mathematician who made awesome contributions to real and complex analysis. Like Dedekind he wished to clarify and put on a firm foundation the basic concepts of analysis.

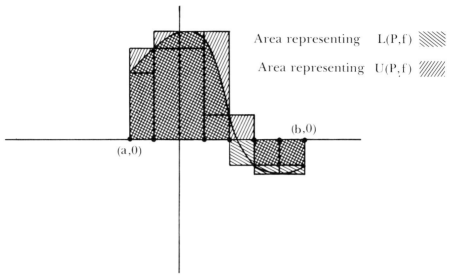

Area representing   L(P,f)

Area representing   U(P,f)

(b,0)

(a,0)

Figure 20

Proposition 1:   If P is any partition of $[a, b]$, then if

$$m = \text{glb } \{f(x) \mid x \in [a, b]\} \quad \text{and} \quad M = \text{lub } \{f(x) \mid x \in [a, b]\},$$

then

$$m(b - a) \leqslant L(P, f) \leqslant U(P, f) \leqslant M(b - a). \tag{6}$$

PROOF: If $P = \{x_0, x_1, \ldots, x_n\}$, then $m \leqslant m_i$ and $M \geqslant M_i$, where m_i and M_i are as defined in (1).   It follows at once that

$$m \sum_{i=0}^{n-1} (x_{i+1} - x_i) = \sum_{i=0}^{n-1} m(x_{i+1} - x_i) \leqslant \sum_{i=0}^{n-1} m_i(x_{i+1} - x_i) \leqslant \sum_{i=0}^{n-1} M_i(x_{i+1} - x_i)$$

$$\leqslant \sum_{i=0}^{n-1} M(x_{i+1} - x_i) = M \sum_{i=0}^{n-1} (x_{i+1} - x_i). \tag{7}$$

But

$$\sum_{i=0}^{n-1} (x_{i+1} - x_i) = (x_1 - x_0) + (x_2 - x_1) + \ldots + (x_{n-1} - x_{n-2}) + (x_n - x_{n-1})$$

$$= x_n + (x_{n-1} - x_{n-1}) + \ldots + (x_1 - x_1) - x_0$$

$$= x_n - x_0$$

$$= b - a.$$

The first and last sums in (7) therefore reduce to $m(b - a)$ and $M(b - a)$, respectively, and the proposition follows immediately.

Corollary: Both $\underline{\int}_a^b f(x)\, dx$ and $\overline{\int}_a^b f(x)\, dx$ exist.
The proof of this corollary is left to the reader.

A partition P may itself be further "subdivided" by adding more points to form a new partition P'. Since P' contains more points than P, (2) and (3) for P' will each contain more terms than they do for P. We now inquire what relationship exists between the Riemann sums relative to a "refinement" P' of P and the corresponding sums for P. We first make the notion of "refinement" more precise.

Definition 3:   *We say that a partition P' of $[a, b]$ is a **refinement** of the partition P, or that P' **refines** P, if $P \subseteq P'$.*

Thus, if P' refines P each end point of an interval relative to P is also the end point of an interval relative to P'. Given two partitions P and P' of $[a, b]$, the partition $P \cup P'$ refines both P and P'; hence any two partitions have a common refinement. It is clear that if P' refines P, then $m(P') \leqslant m(P)$.

Proposition 2:   Suppose that P' is a refinement of P. Then

$$L(P, f) \leqslant L(P', f) \tag{8}$$

and

$$U(P', f) \leqslant U(P, f). \tag{9}$$

Proof: We prove (8) and leave the proof of (9) as an exercise. It suffices to prove (8) for the case that P' contains exactly one more point than P since if Proposition 2 is true for this case, finite induction tells us that Proposition 2 will hold whenever P' contains any finite number of points more than P. Suppose then that $P = \{x_0, x_1, \ldots, x_n\}$ and $P' = P \cup \{y\}$, where $x_i \leqslant y \leqslant x_{i+1}$. Let

$$m_1' = \text{glb } \{f(x) \mid x \in [x_i, y]\} \quad \text{and} \quad m_2' = \text{glb } \{f(x) \mid x \in [y, x_{i+1}]\}.$$

Now $m_1' \geqslant m_i$ and $m_2' \geqslant m_i$, where m_i is as defined in (1). Therefore it follows that

$$\begin{aligned}
L(P', f) - L(P, f) &= m_1'(y - x_i) + m_2'(x_{i+1} - y) - m_i(x_{i+1} - x_i) \\
&= m_1'(y - x_i) + m_2'(x_{i+1} - y) - m_i((x_{i+1} - y) + (y - x_i)) \\
&= (m_1' - m_i)(y - x_i) + (m_2' - m_i)(x_{i+1} - y)
\end{aligned}$$

This last expression, however, is the sum of two non-negative terms, and hence is non-negative. Since $L(P', f) - L(P, f) \geqslant 0$, $L(P', f) \geqslant L(P, f)$.

Corollary: $\underline{\int}_a^b f(x)\, dx \leqslant \overline{\int}_a^b f(x)\, dx.$

PROOF: Let P and P' be any two partitions of $[a, b]$. Then $P \cup P'$ is a common refinement of P and P'. Therefore by Proposition 2, we have

$$L(P, f) \leqslant L(P \cup P', f) \leqslant U(P \cup P', f) \leqslant U(P', f).$$

Therefore $L(P, f) \leqslant U(P', f)$. Since P is arbitrary, it follows that

$$\underline{\int}_a^b f(x)\, dx \leqslant U(P', f). \tag{10}$$

Again, because P' is arbitrary, the corollary follows from (10).

CRITERIA FOR INTEGRABILITY

The following proposition gives an important criterion for a function to be integrable over $[a, b]$.

Proposition 3:   The function f is integrable on $[a, b]$ if and only if given any $p > 0$, there is a partition P such that

$$U(P, f) - L(P, f) < p. \tag{11}$$

PROOF: Suppose first that f is integrable over $[a, b]$. Since $\underline{\int}_a^b f(x)\, dx = \overline{\int}_a^b f(x)\, dx = \int_a^b f(x)\, dx$, there must be partitions P and P' such that

$$U(P, f) - \int_a^b f(x)\, dx < p/2 \tag{12}$$

and

$$\int_a^b f(x)\, dx - L(P', f) < p/2. \tag{13}$$

Now $U(P \cup P', f) \leqslant U(P, f)$ and $L(P \cup P', f) \geqslant L(P', f)$. Therefore (12) and (13) also hold with $U(P \cup P', f)$ replacing $U(P, f)$ and $L(P', f)$. Adding (12) and (13) thus modified, we find $U(P \cup P', f) - L(P \cup P', f) < p$.

Suppose now that given any $p > 0$, there is a partition P such that (11) holds. Since $L(P, f) \leqslant \underline{\int}_a^b f(x)\, dx \leqslant \overline{\int}_a^b f(x)\, dx \leqslant U(P, f)$, we have

$$0 \leqslant \overline{\int}_a^b f(x)\, dx - \underline{\int}_a^b f(x)\, dx < p. \tag{14}$$

Since p was arbitrary, it follows that $\overline{\int}_a^b f(x)\, dx = \underline{\int}_a^b f(x)\, dx$; hence f is integrable over $[a, b]$.

Example 2: Consider the function $f:[0, 1] \to \mathbb{R}$ defined by $f(x) = x$. For each positive integer n, the partition

$$P(n) = \{0, 1/n, 2/n, \ldots, (n-1)/n, n/n = 1\}$$

divides $[0, 1]$ into n equal subintervals. Since f is increasing on $[0, 1]$, we have

$$U(P(n),f) = \sum_{i-1}^{n} (i/n)(1/n) = \sum_{i=1}^{n} i(1/n)^2 = (1/n)^2 \sum_{i=1}^{n} i$$

and

$$L(P(n),f) = \sum_{i=0}^{n-1} (i/n)(1/n) = (1/n)^2 \sum_{i=0}^{n-1} i = (1/n)^2 \left(\sum_{j=1}^{n-1} i \right)$$

(see Example 1). Therefore for any positive integer n,

$$U(P(n),f) - L(P(n),f) = n(1/n)^2 = 1/n. \tag{15}$$

Since $1/n \to 0$, given any $p > 0$, we can find n such that (15) has a value less than p. The hypotheses of Proposition 3 therefore apply to f; hence f is integrable over $[0, 1]$.

Quite similar in statement and proof to Proposition 3 is the following criterion for integrability.

Proposition 4: The function f is integrable over $[a, b]$ if and only if there is a number A such that given any $p > 0$ there is a partition P of $[a, b]$ such that $|U(P,f) - A| < p$ and $|A - L(P,f)| < p$; moreover, $\int_a^b f(x)\, dx = A$.

We leave the proof of Proposition 4 to the reader.

Proposition 5: If f is continuous on $[a, b]$, then f is integrable on $[a, b]$.

PROOF: Let $p > 0$ and p' be such that $p'(b - a) < p$. Since f is continuous on $[a, b]$, a compact set, f is uniformly continuous on $[a, b]$ (Proposition 31 of Chapter 3). Therefore there is $q > 0$ such that if t and t' are points of $[a, b]$ such that $|t - t'| < q$, then $|f(t) - f(t')| < p'$. Let P be any partition of $[a, b]$ with $m(P) < q$. Then for any subinterval $[x_i, x_{i+1}]$ determined by P, $M_i - m_i < p'$. For f actually assumes the values M_i and m_i on $[x_i, x_{i+1}]$ (corollary to Proposition 29, Chapter 3) and since $x_{i+1} - x_i < q$, $M_i - m_i < p'$ follows from the uniform continuity of f and choice of q. Therefore $U(P,f) - L(P,f) = \sum_{i=0}^{n-1} (M_i - m_i)(x_{i+1} - x_i) \leqslant p'(b - a) < p$. Consequently, f is integrable by Proposition 3.

Continuous functions therefore form one important family of integrable functions. Another important class of integrable functions is given in the next proposition.

Proposition 6: If f is either monotone increasing or monotone decreasing on $[a, b]$, then f is integrable.

PROOF: Suppose that f is monotone increasing; the proof for the case that f is monotone decreasing is similar and we leave it to the reader. Let $p > 0$. Choose a positive integer n large enough so that

$$((b - a)/n)(f(b) - f(a)) < p.$$

Let P be a partition of $[a, b]$ of mesh less than $(b - a)/n$. Since f is monotone increasing, for each subinterval $[x_i, x_{i+1}]$ determined by P we have $m_i = f(x_i)$ and $M_i = f(x_{i+1})$. Therefore

$U(P, f) - L(P, f)$

$$= \sum_{i=0}^{n-1} (f(x_{i+1}) - f(x_i))(x_{i+1} - x_i) \leqslant \frac{b - a}{n} \sum_{i=0}^{n-1} (f(x_{i+1}) - f(x_i))$$

$$= \left(\frac{b - a}{n}\right)(f(b) - f(a)) < p.$$

Consequently, f is integrable by Proposition 3.

Since a function need not be continuous to be monotone, we see from Proposition 6 that there are functions which have points of discontinuity in $[a, b]$, but which still are integrable over $[a, b]$. We close this section with an example of a function which is not integrable.

Example 3: Let f be the function defined on $[0, 1]$ defined by

$$f(x) = \begin{cases} 0 & \text{if } x \text{ is rational,} \\ 1 & \text{if } x \text{ is irrational.} \end{cases}$$

For any partition P of $[0, 1]$, we have $U(P, f) = 1$ and $L(P, f) = 0$. Given $0 < p \leqslant 1$, there is no partition P such that $U(P, f) - L(P, f) < p$. Therefore f is not integrable over $[0, 1]$.

EXERCISES

1. Prove the corollary to Proposition 1.

2. Prove (9) of Proposition 2.

3. Prove Proposition 4.

4. Prove Proposition 6 for the case when f is monotone decreasing.

5. In Example 3, prove that for any partition P of $[0, 1]$, $U(P,f) = 1$ and $L(P,f) = 0$.

6. Prove that $\int_a^b f(x)\,dx$ exists if and only if given any $p > 0$, there is $q > 0$ such that for each partition P of $[a, b]$ with mesh less than q, $U(P,f) - L(P,f) < p$.

7. Prove that f is integrable over $[a, b]$ if and only if given any sequence $\{P_n\}$, $n \in \mathbb{N}$, of partitions of $[a, b]$ such that $m(P_n) \to 0$, the sequences $\{U(P_n,f)\}$, $n \in \mathbb{N}$, and $\{L(P_n,f)\}$, $n \in \mathbb{N}$, both converge to the same limit.

8. Use Proposition 7 to find $\int_0^1 f(x)\,dx$, where $f(x) = x$. Since f is continuous on $[0, 1]$, we know that this integral exists. Let P_n be the partition which divides $[0, 1]$ into n equal subintervals. Compute the limit of the sequence $\{U(P_n,f)\}$, $n \in \mathbb{N}$.

6.2 MORE CRITERIA FOR INTEGRABILITY

RIEMANN SUMS

Suppose $P = \{x_0, x_1, \ldots, x_n\}$ is a partition of $[a, b]$ and f is a bounded function defined on $[a, b]$. If $t_i \in [x_i, x_{i+1}]$, $i = 0, \ldots, n - 1$, then

$$L(P,f) = \sum_{i=0}^{n-1} m_i(x_{i+1} - x_i) \leqslant \sum_{i=0}^{n-1} f(t_i)(x_{i+1} - x_i) \leqslant \sum_{i=0}^{n-1} M_i(x_{i+1} - x_i)$$
$$= U(P,f) \quad \textbf{(16)}$$

If f is integrable over $[a, b]$, then as we take finer and finer partitions of $[a, b]$, $L(P,f)$ and $U(P,f)$ will both "converge" to $\int_a^b f(x)\,dx$. Because of the inequalities expressed in (16), it would appear that $\sum_{i=0}^{n-1} f(t_i)(x_{i+1} - x_i)$ should also "converge" to $\int_a^b f(x)\,dx$. However intuitively appealing the previous statement may have sounded, the following discussion will point up how much work is necessary to make that statement precise and then to prove it.

Definition 4: *The sum $\sum_{i=0}^{n-1} f(t_i)(x_{i+1} - x_i)$ appearing in (16) is said to be a **Riemann sum of f over [a, b] relative to P.***

Note that the t_i are arbitrary in that t_i can be any point whatsoever of $[x_{i+1}, x_i]$. We now define what we mean by convergence of Riemann sums.

Definition 5: *We continue to assume (as we will throughout this chapter) that f is a bounded function on* $[a, b]$. *We say that the Riemann sums of f on* $[a, b]$ **converge** *to some number L if given any* $p > 0$, *there is* $q > 0$ *such that if P is any partition of mesh less than q, then*

$$|S(P, f) - L| < p,$$

where $S(P, f)$ *is any Riemann sum of f over* $[a, b]$ *relative to P.*

We would like to be able to say that f is integrable over $[a, b]$ with $\int_a^b f(x)\, dx = L$ if and only if the Riemann sums of f over $[a, b]$ converge to L. We prove this statement in the next two propositions.

Proposition 7: If the Riemann sums of f over $[a, b]$ converge to L, then f is integrable over $[a, b]$ with $\int_a^b f(x)\, dx = L$.

PROOF: The proof will use Proposition 4 of the previous section. We begin by proving that if P is any partition of $[a, b]$ and $p > 0$, then there is a Riemann sum $S(P, f)$ of f over $[a, b]$ relative to P such that

$$|U(P, f) - S(P, f)| < p/2.$$

As usual we let the points of P be $x_0, x_1, \ldots, x_n$. Since

$$M_i = \text{lub } \{f(x) \mid x \in [x_i, x_{i+1}]\},$$

there is $t_i \in [x_i, x_{i+1}]$ such that

$$0 \leqslant M_i - f(t_i) < p/(2(b - a)). \tag{17}$$

Consequently,

$$0 \leqslant M_i(x_{i+1} - x_i) - f(t_i)(x_{i+1} - x_i) < (p/(2(b - a)))(x_{i+1} - x_i). \tag{18}$$

It follows from (18) that

$$0 \leqslant \sum_{i=0}^{n-1} M_i(x_{i+1} - x_i) - \sum_{i=0}^{n-1} f(t_i)(x_{i+1} - x_i)$$

$$= U(P, f) - \sum_{i=0}^{n-1} f(t_i)(x_{i+1} - x_i) < (p/(2(b - a)))(b - a) = p/2. \tag{19}$$

Letting $S(P, f) = \sum_{i=0}^{n-1} f(t_i)(x_{i+1} - x_i)$, we have $|U(P, f) - S(P, f)| < p/2$.

Similarly, we can prove that given any partition P of $[a, b]$ and any $p > 0$, there is a Riemann sum $S'(P, f)$ of f over $[a, b]$ relative to P such that $|S'(P, f) - L(P, f)| < p/2$.

Let $p > 0$. Since the Riemann sums converge to L, there is $q > 0$ such that if P is any partition of $[a, b]$ with $m(P) < q$, then $|S(P, f) - L| < p/2$, where $S(P, f)$ is any Riemann sum of f over $[a, b]$ relative to P. Let P be a partition with $m(P) < q$. By what was proved earlier, there are Riemann sums $S(P, f)$ and $S'(P, f)$ such that $|U(P, f) - S(P, f)| < p/2$ and $|L(P, f) - S'(P, f)| < p/2$. Moreover, since $m(P) < q$, we also have $|S(P, f) - L| < p/2$ and $|S'(P, f) - L| < p/2$. Therefore

$$|U(P, f) - L| = |(U(P, f) - S(P, f)) + (S(P, f) - L)|$$
$$\leqslant |U(P, f) - S(P, f)| + |S(P, f) - L| < p/2 + p/2 = p. \quad (20)$$

Similarly, we have

$$|L(P, f) - L| = |(L(P, f) - S'(P, f)) + (S'(P, f) - L)| < p. \quad (21)$$

Therefore Proposition 7 follows from Proposition 4.

Proposition 8: If f is integrable over $[a, b]$, then the Riemann sums of f over $[a, b]$ converge to $\int_a^b f(x)\, dx$.

PROOF: Set $\int_a^b f(x)\, dx = L$. Let $p > 0$. From Proposition 4 and the corollary to Proposition 2 we see that there is a partition $P = \{x_0, x_1, \ldots, x_n\}$ of $[a, b]$ such that

$$U(P, f) - L < p/2 \quad \text{and} \quad L - L(P, f) < p/2. \quad (22)$$

Since f is bounded on $[a, b]$, there is a positive number M such that $|f(x)| \leqslant M$ for all $x \in [a, b]$. Set

$$q = p/(4Mn). \quad (23)$$

Let P' be any partition of mesh less than q and consider $P \cup P'$. We can consider $P \cup P'$ as being formed by the addition of at most $n - 1$ points to P' (since P contains $n + 1$ points, but two of these are a and b). If the subinterval $[y_{i+1}, y_i]$ determined by P' is divided by the addition of points of P into m subintervals, none of these subintervals can contribute more to $U(P, f) - U(P \cup P', f)$ than $2Mq$. Hence, since adding the points of P to P' adds no more than $n - 1$ more subintervals to the intervals already determined by P', we find that

$$U(P, f) - U(P \cup P', f) \leqslant (n - 1)2Mq = (p/(4Mn))(n - 1)2M < p \quad (24)$$

and, similarly,

$$L(P \cup P', f) - L(P, f) \leqslant (n - 1)2Mq < p. \quad (25)$$

It follows from (22), (24), and (25) and the facts $L(P',f) \leqslant L(P \cup P',f)$ and $U(P \cup P',f) \leqslant U(P', f)$ that

$$U(P',f) - L < p \quad \text{and} \quad L - L(P',f) < p \tag{26}$$

if P' is any partition with $m(P') < q$.

Now if P'' is any partition of mesh less than q and $S(P'',f)$ is any Riemann sum relative to P'', then $L(P'',f) \leqslant S(P'',f) \leqslant U(P'',f)$ (see (16)). But this together with (26) implies that $|S(P'',f) - L| < p$. Therefore the Riemann sums converge to $L = \int_a^b f(x)\, dx$.

Propositions 7 and 8 together give us the following proposition.

Proposition 9: f is integrable over $[a, b]$ if and only if the Riemann sums of f over $[a, b]$ converge to some limit L. Moreover, $\int_a^b f(x)\, dx = L$.

Proposition 10: $\int_a^b f(x)\, dx = L$ if and only if given any sequence $\{P_n\}$, $n \in \mathbb{N}$, of partitions of $[a, b]$ such that $m(P_n) \to 0$, any sequence $\{S(P_n,f)\}$, $n \in \mathbb{N}$, of Riemann sums relative to the P_n converges to L.

PROOF: Suppose first that $\int_a^b f(x)\, dx = L$ and $P_1, P_2, \ldots$ is a sequence of partitions of $[a, b]$ with $m(P_n) \to 0$. Let $p > 0$. Since $\int_a^b f(x)\, dx = L$, there is $q > 0$ such that for any partition P of $[a, b]$ $m(P) < q$ implies $|S(P,f) - L| < p$, where $S(P,f)$ is any Riemann sum relative to P. Since $m(P_n) \to 0$, there is an integer M such that $n > M$ implies $m(P_n) < q$. Therefore if $n > M$, $|S(P_n,f) - L| < p$. Consequently, $S(P_n,f) \to L$.

Suppose now that $\int_a^b f(x)\, dx \neq L$. Then it is possible to find $p > 0$ for which there is no $q > 0$ such that $m(P) < q$ implies $|S(P,f) - L| < p$, P a partition and $S(P,f)$ a Riemann sum relative to P. Set $q_n = 1/n$. For each positive integer n, let P_n be a partition of mesh at most q_n and $S(P_n,f)$ a Riemann sum relative to P_n such that $|S(P_n,f) - L| > p$. Then the sequence $\{P_n\}$, $n \in \mathbb{N}$, of partitions thus obtained has $m(P_n) \to 0$, but $S(P_n,f)$ does not converge to L (since no $S(P_n,f)$ is in $N(L, p)$). This completes the proof of Proposition 10.

Proposition 10 enables us to consider "nice" partitions and Riemann sums to compute an integral provided that we know that the integral exists. The following example illustrates this point.

Example 4: Consider $f(x) = 3x + 1$ over the interval $[1, 2]$. Since f is continuous on $[1, 2]$, $\int_1^2 f(x)\, dx$ exists. To compute $\int_1^2 f(x)\, dx$ we can use any sequence of partitions with mesh converging to 0 and any sequence of Riemann sums relative to the partitions of this sequence. In particular, we let P_n be the partition which divides $[1, 2]$ into n equal subintervals. Each subinterval will have length $(2 - 1)/n = 1/n$. Therefore the points

x_i, $i = 0, \ldots, n$, of P_n will be $1, 1 + 1/n, 1 + 2/n, \ldots, 1 + (n - 1)/n$, $1 + n/n = 2$. We let $t_i = x_i$, $i = 0, \ldots, n - 1$. Then*

$$S(P_n, f) = \sum_{i=0}^{n-1} f(t_i)(x_{i+1} - x_i)$$

$$= \sum_{i=0}^{n-1} (3(1 + i/n) + 1)(1/n) = 3\left(\frac{n-1}{n}\right) + \frac{3}{n^2}\left(\sum_{i=0}^{n-1} i\right) + \frac{n-1}{n}$$

$$= 4\left(\frac{n-1}{n}\right) + (3/n^2)((n-1)/2)n. \tag{27}$$

The limit of (27) is $4 + 3/2 = 11/2$.

Naturally, even the process carried out in Example 4, however simple it may be in a relative sense, is still too cumbersome for practical use in evaluating integrals (although it might be quite successful in a modified form in estimating integrals using computer techniques). The next section will deal with certain propositions which do help in finding many integrals. We must point out though that in concerning ourselves as we are in this text with the theory of the real numbers and the calculus, we are not so much interested in finding integrals as in rigorously defining the notion of an integral, determining under what conditions an integral exists, and proving some of the fundamental properties of integrals.

EXERCISES

1. In the proof of Proposition 7, prove that given any partition P of $[a, b]$ and any $p > 0$, there is a Riemann sum $S'(P, f)$ of f over $[a, b]$ relative to P such that

$$|S'(P, f) - L(P, f)| < p/2.$$

2. Prove that if $f(x) = k$, k a constant, for all $x \in [a, b]$, then $\int_a^b f(x)\, dx = k(b - a)$.

3. We have at least implicitly assumed that the closed interval $[a, b]$ is non-degenerate, that is, $a < b$. Prove that if $a = b$, then $\int_a^b f(x)\, dx = 0$.

4. A function f on $[a, b]$ is said to be a *step function* if there are points $a = x_0 < x_1 < x_2 < \ldots < x_n = b$ such that

---

* We have used the well-known formula for the sum of the first n positive integers: $\sum_{i=1}^{n} i = (i/2)(i + 1)$. See Example 5, Chapter 1.

f is constant on each of the open intervals $(x_0, x_1), \ldots, (x_{n-1}, x_n)$. Prove that if f is a step function on $[a, b]$, then f is integrable over $[a, b]$.

5. Compute $\int_{-1}^{1} |x|\, dx$ directly in the manner of Example 4.

6. Prove that if $f(x)$ is non-negative for each $x \in [a, b]$, then if $\int_a^b f(x)\, dx$ exists, it is non-negative.

7. Prove that f is integrable over $[a, b]$ if and only if given any $p > 0$, there is $q > 0$ such that if P and P' are any partitions of $[a, b]$ of mesh less than q, then

$$|S(P, f) - S(P', f)| < p,$$

where $S(P, f)$ and $S(P', f)$ are any Riemann sums of f over $[a, b]$ (relative to P and P'). (This is the analog for Riemann sums of the Cauchy property for sequences.)

6.3   MORE BASIC PROPERTIES OF THE RIEMANN INTEGRAL

Proposition 11:   If $\int_a^b f(x)\, dx = L$ and r is any real number, then $\int_a^b (rf)(x)\, dx = rL$.

PROOF: We note that for any partition P of $[a, b]$, $S(P, rf) = rS(P, f)$, where $S(P, f)$ is any Riemann sum. Therefore if P_n is a sequence of partitions with $m(P_n) \to 0$, then $\{S(P_n, rf)\}$, $n \in \mathbb{N}$, which is the same as $\{rS(P_n, f)\}$, $n \in \mathbb{N}$, converges to rL (Proposition 2(b) of Chapter 4). Therefore $\int_a^b (rf)(x)\, dx = rL$ by Proposition 10.

Proposition 12:   If $\int_a^b f(x)\, dx = L$ and $\int_a^b g(x)\, dx = L'$, then

$$\int_a^b (f + g)(x)\, dx \text{ exists and is equal to } L + L'.$$

PROOF: Proposition 12 follows from Proposition 10 of this chapter and Proposition 2 of Chapter 4. We leave the details as an exercise for the reader.

Combining Propositions 11 and 12, we have the following corollary.

Corollary: If $\int_a^b f(x)\, dx = L$ and $\int_a^b g(x)\, dx = L'$, then $\int_a^b (rf + sg)(x)\, dx$ exists and is equal to $rL + sL'$, where r and s are any real numbers.

Proposition 13:   If f and g are both integrable on $[a, b]$ and if $f(x) \geqslant g(x)$ for all $x \in [a, b]$, then $\int_a^b f(x)\, dx \geqslant \int_a^b g(x)\, dx$.

PROOF: Since $f(x) \geqslant g(x)$ for all $x \in [a, b]$, $f(x) - g(x) = (f - g)(x) = (f + (-1)g)(x) \geqslant 0$ for all $x \in [a, b]$; then $f - g$ is integrable on $[a, b]$ by the preceding corollary. Therefore

$$\int_a^b (f - g)(x) \, dx = \int_a^b f(x) \, dx - \int_a^b g(x) \, dx \geqslant 0$$

(Exercise 6 of the preceding section). The result follows at once.

Proposition 14:   If f is integrable over $[a, b]$, then $|f|$ (defined by $|f|(x) = |f(x)|$ for all $x \in [a, b]$) is also integrable over $[a, b]$. Moreover,

$$\left| \int_a^b f(x) \, dx \right| \leqslant \int_a^b |f|(x) \, dx. \tag{28}$$

PROOF: We first show that $|f|$ is integrable over $[a, b]$. Since f is integrable over $[a, b]$, given any $p > 0$, we can find a partition P such that $U(P, f) - L(P, f) < p$. From Proposition 1(e) of Chapter 3 we see that

$$||f(t)| - |f(t')|| \leqslant |f(t) - f(t')|, \tag{29}$$

for any t and t' in $[a, b]$. Consequently, $M_i' - m_i' \leqslant M_i - m_i$, where M_i and m_i have the meaning assigned in (1) relative to f and M_i' and m_i' are the corresponding numbers for $|f|$. Therefore

$$U(P, |f|) - L(P, |f|) = \sum_{i=0}^{n-1} (M_i' - m_i')(x_{i+1} - x_i)$$

$$\leqslant \sum_{i=0}^{n-1} (M_i - m_i)(x_{i+1} - x_i)$$

$$= U(P, f) - L(P, f) < p.$$

Thus, Proposition 3 $|f|$ is integrable over $[a, b]$.

Since $f(x) \leqslant |f(x)| = |f|(x)$ and $-f(x) \leqslant |f(x)| = |f|(x)$ for all x in $[a, b]$, by Proposition 13 we have

$$\int_a^b f(x) \, dx \leqslant \int_a^b |f|(x) \, dx \tag{30}$$

and

$$-\int_a^b f(x) \, dx = \int_a^b (-f)(x) \, dx \leqslant \int_a^b |f|(x) \, dx. \tag{31}$$

Because $|\int_a^b f(x) \, dx|$ is either $\int_a^b f(x) \, dx$ or $-\int_a^b f(x) \, dx$, (30) and (31) imply

$$\left| \int_a^b f(x) \, dx \right| \leqslant \int_a^b |f|(x) \, dx,$$

which is what we wanted to prove.

Proposition 15: If f is integrable over $[a, b]$, then f^2 is also integrable over $[a, b]$.

PROOF: Since f is bounded on $[a, b]$, $|f|$ is also bounded on $[a, b]$. Since $|f|$ is integrable over $[a, b]$ (because f is integrable), given any $p > 0$ there is a partition P of $[a, b]$ such that

$$U(P, |f|) - L(P, |f|) < p/2M, \tag{32}$$

where $|f(x)| \leqslant M$ for all x in $[a, b]$. For this partition P let M_i and m_i be as defined in (1) for the function $|f|$ and let M'_i and m'_i be the corresponding quantities for f^2. Then we have

$$U(P, f^2) - L(P, f^2) = \sum_{i=0}^{n-1} (M'_i - m'_i)(x_{i+1} - x_i)$$

$$= \sum_{i=0}^{n-1} (M_i^2 - m_i^2)(x_{i+1} - x_i)$$

$$= \sum_{i=0}^{n-1} (M_i - m_i)(M_i + m_i)(x_{i+1} - x_i)$$

$$\leqslant 2M \left(\sum_{i=0}^{n-1} (M_i - m_i)(x_{i+1} - x_i) \right)$$

$$= 2M(U(P, |f|) - L(P, |f|)) < 2Mp/(2M) = p.$$

Therefore f^2 is integrable over $[a, b]$ by Proposition 3.

Corollary: If f and g are both integrable over $[a, b]$, then fg is also integrable over $[a, b]$.

PROOF: Since f and g are integrable over $[a, b]$, f^2, g^2, and $(f + g)^2$ are all integrable over $[a, b]$. Since $fg = (1/2)((f + g)^2 - f^2 - g^2)$, fg is also integrable over $[a, b]$.

Proposition 16: If f is integrable over $[a, c]$ and if $a < b < c$, then f is integrable over both $[a, b]$ and $[b, c]$.

PROOF: Since f is integrable over $[a, c]$, given $p > 0$ there is a partition P of $[a, c]$ such that $U(P, f) - L(P, f) < p$. Let P' be the partition formed by adding the point b to P. Then P' is a refinement of P, hence

$$U(P, f) \geqslant U(P', f) \geqslant L(P', f) \geqslant L(P, f) \tag{33}$$

by Proposition 2. It follows at once that

$$U(P', f) - L(P', f) < p. \tag{34}$$

Now since b is a point of P', we can consider P' as $P_1 \cup P_2$, where P_1 is the set of points of P' between a and b, inclusive, and P_2 is the set of points of P' between b and c, inclusive. Therefore

$$U(P', f) = U(P_1, f) + U(P_2, f) \quad \text{and} \quad L(P', f) = L(P_1, f) + L(P_2, f).$$
(35)

Hence

$$U(P', f) - L(P', f) = (U(P_1, f) - L(P_1, f)) + (U(P_2, f) - L(P_2, f)) < p.$$

But since

$$U(P_1, f) - L(P_1, f) \quad \text{and} \quad U(P_2, f) - L(P_2, f) \qquad (36)$$

are both non-negative, it follows that both of the quantities in (36) are less than p. Therefore f is integrable over both $[a, b]$ and $[b, c]$.

Proposition 17:   If f is integrable over $[a, c]$ and $a < b < c$, then f is integrable over $[a, b]$ and $[b, c]$ (Proposition 16), and

$$\int_a^b f(x)\ dx + \int_b^c f(x)\ dx = \int_a^c f(x)\ dx.$$

PROOF: Let $p > 0$. Since f is integrable over both $[a, b]$ and $[b, c]$, there are partitions P_1 and P_2 of $[a, b]$ and $[b, c]$, respectively, such that

$$\left| U(P_1, f) - \int_a^b f(x)\ dx \right| < p/2 \quad \text{and} \quad \left| \int_a^b f(x)\ dx - L(P_1, f) \right| < p/2 \quad (37)$$

and

$$\left| U(P_2, f) - \int_b^c f(x)\ dx \right| < p/2 \quad \text{and} \quad \left| \int_b^c f(x)\ dx - L(P_2, f) \right| < p/2. \quad (38)$$

Let $P = P_1 \cup P_2$. Then P is a partition of $[a, c]$; moreover, $U(P, f) = U(P_1, f) + U(P_2, f)$ and $L(P, f) = L(P_1, f) + L(P_2, f)$. It therefore follows from (37) and (38) that

$$\left| U(P, f) - \left(\int_a^b f(x)\ dx + \int_b^c f(x)\ dx \right) \right| < p \qquad (39)$$

and

$$\left| \left(\int_a^b f(x)\ dx + \int_b^c f(x)\ dx \right) - L(P, f) \right| < p. \qquad (40)$$

Proposition 18 follows from (39) and (40) and Proposition 4.

The following proposition is the converse of Propositions 16 and 17. We leave its proof to the reader.

Proposition 18: Suppose that $a < b < c$ and that f is integrable over both $[a, b]$ and $[b, c]$. Then f is integrable over $[a, c]$ and

$$\int_a^c f(x) \, dx = \int_a^b f(x) \, dx + \int_b^c f(x) \, dx.$$

Corollary: If f is bounded and has but a finite number of discontinuities in $[a, b]$, then f is integrable over $[a, b]$.

We also leave the proof of this corollary to the reader.

We have defined what we mean by $\int_a^b f(x) \, dx$, where $a < b$. We now extend this definition slightly.

Definition 6: *If f is integrable over $[a, b]$, we define*

$$\int_b^a f(x) \, dx = -\int_a^b f(x) \, dx. \tag{41}$$

Definition 6 in turn enables us to extend Proposition 18.

Proposition 19: If a, b, and c are any real numbers and f is integrable over the closed interval which contains a, b, and c, then

$$\int_a^b f(x) \, dx + \int_b^c f(x) \, dx = \int_a^c f(x) \, dx.$$

Note that in Proposition 19 we do not make any assumptions about the order of a, b, and c. We leave the proof of Proposition 19 to the reader.

EXERCISES

1. Prove Proposition 18.

2. Prove the corollary to Proposition 18.

3. Prove Proposition 19.

4. Let $f(x) = (x + 1)(x - 1)x$, $x \in [-2, 2]$. Set $f^+ = (1/2)(|f| + f)$ and $f^- = (1/2)(|f| - f)$. Graph f, f^+, and f^- together. Indicate how f^+ and f^- are related to f and $|f|$. Prove that f^+ and f^- are integrable over $[-2, 2]$. Generalize to a function f integrable on $[a, b]$.

5. Prove that if f and g are integrable over $[a, b]$, then $f^n g^m$ is also integrable over $[a, b]$, where m and n are any two positive integers.

6. Suppose f is integrable over $[a, b]$. Under what conditions is $1/f$ integrable over $[a, b]$? State and prove a theorem about the integrability of $1/f$ over $[a, b]$.

7. Suppose f is a polynomial function on $[a, b]$. Prove directly from the propositions of this section (rather than appealing to the continuity of f) that f is integrable over $[a, b]$.

6.4 THE FUNDAMENTAL THEOREM OF THE CALCULUS

We now rigorously prove a statement that is one of the first things taught students beginning integral calculus: If F is a differentiable on $[a, b]$ with $F'(x) = f(x)$ for each $x \in [a, b]$ and if f is continuous on $[a, b]$, then $\int_a^b f(x)\, dx = F(b) - F(a)$. We will also investigate one other form of this basic theorem as well as prove some of its consequences.

Proposition 20 (*Fundamental Theorem of the Calculus, First Form*):   If f is integrable on $[a, b]$ and there exists a function F defined and differentiable on $[a, b]$ such that $F'(x) = f(x)$, then $\int_a^b f(x)\, dx = F(b) - F(a)$. (Note that the condition that f is integrable over $[a, b]$ is actually weaker than the condition that f is continuous on $[a, b]$.)

PROOF: Let $P = \{x_0, x_1, \ldots, x_n\}$ be any partition of $[a, b]$. By Proposition 17 of Chapter 5 there is $t_i \in [x_{i+1}, x_i]$ such that

$$F(x_{i+1}) - F(x_i) = f(t_i)(x_{i+1} - x_i). \tag{42}$$

Therefore

$$\sum_{i=0}^{n-1} f(t_i)(x_{i+1} - x_i) = \sum_{i=0}^{n-1} (F(x_{i+1}) - F(x_i)) = F(b) - F(a). \tag{43}$$

The left side of (43) is a Riemann sum of f over $[a, b]$. This sum will approach $\int_a^b f(x)\, dx$ as its limit as the mesh of P approaches 0. Since P was arbitrary, it follows from (43) that $\int_a^b f(x)\, dx = F(b) - F(a)$.

Proposition 21 (*Fundamental Theorem of the Calculus, Second Form*):   Suppose that f is integrable over $[a, b]$. For each $t \in [a, b]$, set

$$F(t) = \int_a^t f(x)\, dx. \tag{44}$$

(The integral in (44) exists by Proposition 16.) Then F is continuous on $[a, b]$. Moreover, if f is continuous at $t_0 \in [a, b]$, then $F'(t_0)$ exists and is equal to $f(t_0)$.

PROOF: Since f is a bounded function on $[a, b]$ (we have only defined the integral for bounded functions), there is a positive number M such that $|f(x)| \leq M$ for all $x \in [a, b]$. Let $p > 0$, and set $q < p/M$. Then if t and t' are two points of $[a, b]$ such that $|t - t'| < q$, then

$$
\begin{aligned}
|F(t) - F(t')| &= \left| \int_a^t f(x)\, dx - \int_a^{t'} f(x)\, dx \right| \\
&= \left| \int_a^t f(x)\, dx + \int_{t'}^a f(x)\, dx \right| \\
&= \left| \int_{t'}^t f(x)\, dx \right| \leqslant M\,|t - t'| < Mp/M = p.
\end{aligned}
$$

Consequently, F is continuous on $[a, b]$.

We now assume that f is continuous at t_0. Therefore given $p > 0$ we can find $q > 0$ such that if $t \in [a, b]$ and $|t - t_0| < q$, then $|f(t) - f(t_0)| < p$. It follows then that $t_0 < t < t_0 + q$ with $t \in [a, b]$ implies

$$
\begin{aligned}
\left| \frac{F(t) - F(t_0)}{t - t_0} - f(t_0) \right| &= \frac{\left| \int_{t_0}^t (f(x) - f(t_0))\, dx \right|}{t - t_0} \\
&\leqslant p(t - t_0)/(t - t_0) = p.
\end{aligned}
\tag{45}
$$

Similarly, if $t \in [a, b]$ and $t_0 - q < t < t_0$, then

$$
\left| \frac{F(t) - F(t_0)}{t - t_0} - f(t_0) \right| = \frac{\left| \int_{t_0}^t (f(x) - f(t_0))\, dx \right|}{t_0 - t} \leqslant p.
\tag{46}
$$

Together (45) and (46) imply

$$
\lim_{t \to t_0} \frac{F(t) - F(t_0)}{t - t_0} = F'(t_0) = f(t_0),
$$

which is what we wanted to prove.

Using Proposition 21 we can prove the "weak," or most standard, form of Proposition 20.

Corollary: If $f : [a, b] \to \mathbb{R}$ is continuous, then there is a function F such that $F'(x) = f(x)$ for each $x \in [a, b]$. Moreover, if F is such a function, then $\int_a^b f(x)\, dx = F(b) - F(a)$.

PROOF: This corollary is simply (44) with $t = b$.

Not only are Propositions 20 and 21 valuable from a computational point of view (in that they enable us to compute many Riemann integrals without going through the lengthy process of computing the limit of Riemann sums), but these propositions also yield other important properties of integrals.

Proposition 22 (*Change of Variable Theorem for Integrals*):   Suppose that g is a differentiable function on $[a, b]$ and f is continuous on the interval $g([a, b])$ (this is an interval by the corollary of Proposition 35 in Chapter 3). Then

$$\int_a^b f(g(x))g'(x)\, dx = \int_{g(a)}^{g(b)} f(t)\, dt. \tag{47}$$

PROOF: Assume first that $g(a) \leqslant g(b)$. Then if F is a function differentiable on $[g(a), g(b)]$ such that $F'(x) = f(x)$ for each $x \in [g(a), g(b)]$, then $\int_{g(a)}^{g(b)} f(t)\, dt = F(g(b)) - F(g(a))$.   Such an F exists by Proposition 21; specifically, $F(x) = \int_{g(a)}^x f(t)\, dt$.   Set $G(x) = F(g(x))$ for each $x \in [a, b]$. Then $G'(x) = F'(g(x))g'(x)$ (by the Chain Rule) $= f(g(x))g'(x)$, for $x \in [a, b]$. Consequently, $\int_a^b f(g(x))g'(x)\, dx = G(b) - G(a) = F(g(b)) - F(g(a))$. Since both sides of (47) are equal to $F(g(b)) - F(g(a))$, they are equal to each other.
     If $g(b) < g(a)$, then

$$\int_{g(a)}^{g(b)} f(t)\, dt = -\int_{g(b)}^{g(a)} f(t)\, dt = -(F(g(a)) - F(g(b))) = F(g(b)) - F(g(a))$$

and the conclusion follows as before.

Proposition 23 (*Integration by Parts*):   If f and g are both differentiable on $[a, b]$ and if f' and g' are both integrable on $[a, b]$, then

$$\int_a^b f(x)g'(x)\, dx = (f(b)g(b) - f(a)g(a)) - \int_a^b g(x)f'(x)\, dx. \tag{48}$$

PROOF: All the integrals in (48) exist. For f and g are differentiable, and hence continuous on $[a, b]$; therefore f and g are integrable over $[a, b]$. Since f, g, f', and g' are all integrable over $[a, b]$, fg' and gf' are integrable over $[a, b]$ (the corollary to Proposition 15). Set $F(x) = f(x)g(x)$ for all $x \in [a, b]$. Then $F'(x) = f(x)g'(x) + g(x)f'(x)$. Consequently,

$$\int_a^b F'(x)\, dx = F(b) - F(a)$$

$$= \int_a^b (f(x)g'(x) + g(x)f'(x))\, dx$$

$$= \int_a^b f(x)g'(x)\, dx + \int_a^b g(x)f'(x)\, dx.$$

Since $F(b) - F(a) = f(b)g(b) - f(a)g(a)$, we obtain the desired result from $F(b) - F(a) = \int_a^b f(x)g'(x)\,dx + \int_a^b g(x)f'(x)\,dx$.

We have two Mean Value Theorems for derivatives. We also have two Mean Value Theorems for integrals.

Proposition 24 (*First Mean Value Theorem for Integrals*): If f is continuous on $[a, b]$, then there is a point x' in $[a, b]$ such that

$$\int_a^b f(x)\,dx = f(x')(b - a).$$

PROOF: Let $M = \text{lub}\{f(x) \mid x \in [a, b]\}$ and $m = \text{glb}\{f(x) \mid x \in [a, b]\}$. Then from Proposition 1 we have

$$m(b - a) \leqslant \int_a^b f(x)\,dx \leqslant M(b - a). \tag{49}$$

Therefore there is a number $c \in [m, M]$ such that

$$\int_a^b f(x)\,dx = c(b - a). \tag{50}$$

Since f is continuous and $[a, b]$ is compact, m and M are actually function values of f on $[a, b]$. Therefore by the Intermediate Value Theorem, there is $x' \in [a, b]$ such that $f(x') = c$. Substituting $f(x')$ for c in (50) we have the desired result.

Proposition 25 (*Second Mean Value Theorem for Integrals*):   Suppose that f and g are both continuous on $[a, b]$ and $g(x) \geqslant 0$ for all $x \in [a, b]$. Then there is $x' \in [a, b]$ such that

$$\int_a^b f(x)g(x)\,dx = f(x') \int_a^b g(x)\,dx. \tag{51}$$

PROOF: Let M and m be the maximum and minimum values of $f(x)$ on the compact set $[a, b]$. From the fact that g is non-negative on $[a, b]$, it follows that

$$m \int_a^b g(x)\,dx \leqslant \int_a^b f(x)g(x)\,dx \leqslant M \int_a^b g(x)\,dx. \tag{52}$$

Since Proposition 25 is trivially true if $g(x) = 0$ for all $x \in [a, b]$, we may assume that $g(x) > 0$ for some $x \in [a, b]$. Therefore one can prove (the proof is left as an exercise) that $\int_a^b g(x)\,dx > 0$. Dividing all parts of (52) by

$\int_a^b g(x)\, dx$ (which is non-zero since it is positive), we obtain

$$m \leqslant \frac{\displaystyle\int_a^b f(x)g(x)\, dx}{\displaystyle\int_a^b g(x)\, dx} \leqslant M. \tag{53}$$

But since m and M are function values of f and f is continuous, there is $x' \in [a, b]$ such that $f(x')$ is the middle term of (53). Expression (51) follows at once.

EXERCISES

1. Suppose g is continuous on $[a, b]$ and $g(x) > 0$ for all $x \in [a, b]$. Prove that $\int_a^b g(x)\, dx > 0$.

2. Prove the series of inequalities (52) in the proof of Proposition 25.

3. Can the point x' of Proposition 24 always be found in (a, b)? Find a function f defined on some closed interval which satisfies the hypotheses of Proposition 24, but for which x' must be an end point of the interval.

4. We have proved an Intermediate Value Theorem for Derivatives (cf. Proposition 23 of Chapter 5). Can we therefore say: If f is the derivative of some function on $[a, b]$, then there is a point x' in $[a, b]$ such that $\int_a^b f(x)\, dx = f(x')(b - a)$? State precisely the reasons for your answer.

5. Suppose f is continuous on $[a, b]$ and $F(x) = \int_a^x f(t)\, dt$ for each $x \in [a, b]$. Does the First Mean Value Theorem for Derivatives apply to F over $[a, b]$? State this theorem for F. State the First Mean Value Theorem for Integrals to $\int_a^b f(x)\, dx$. Compare the two statements. Try to relate the Second Mean Value Theorems for derivatives and integrals.

6. Suppose that f is continuous and monotone increasing and g is continuous on $[a, b]$. Prove that there is $c \in (a, b)$ such that $\int_a^b (fg)(x)\, dx = f(a) \int_a^c g(x)\, dx + f(b) \int_c^b g(x)\, dx$.

7. Suppose f and g are both integrable on $[a, b]$ and $\int_a^b f(x)\, dx = \int_a^b g(x)\, dx$. Prove that $f(c) = g(c)$ for some $c \in [a, b]$.

6.5 EXTENSIONS OF THE INTEGRAL

IMPROPER INTEGRALS

In the previous sections we defined and explored some of the basic properties of $\int_a^b f(x)\,dx$, where f is an integrable function on $[a, b]$. In order for $\int_a^b f(x)\,dx$ to exist, it is necessary that f be bounded on $[a, b]$.

Suppose now that f is a function defined on an open interval (a, b) such that given any $p > 0$, f is integrable on the closed interval $[a + p, b - p]$. Then for any $p > 0$,

$$\int_{a+p}^{b-p} f(x)\,dx \tag{54}$$

is a function of p; hence we may talk about the limit, if one exists, of (54) as p approaches 0.

Definition 7: *If f is a function defined on (a, b) and integrable on $[a + p, b - p]$ for any $p > 0$ (such that $a + p < b - p$), then if (54) approaches a limit L as $p \to 0$, we call L the **improper integral** of f over (a, b); the improper integral of f over (a, b) is denoted by $\int_a^b f(x)\,dx$. If the improper integral $\int_a^b f(x)\,dx$ exists, we say that the integral **converges;** if this improper integral does not exist, we say that the integral **diverges.***

Example 5: Consider the function $f(x) = x^2$ defined on the open interval $(0, 1)$. Then for each $p > 0$ $(p < 1/2)$, we have

$$\int_p^{1-p} x^2\,dx = (1 - p)^3/3 - p^3/3.$$

Since $\lim_{p \to 0} ((1 - p)^3/3 - p^3/3) = 1/3$, it follows that $\int_0^1 x^2\,dx$ converges and is equal to $1/3$.

We will now define a notion of integration over intervals which are not bounded, that is, which have ∞ or $-\infty$ as one of their end points.

Definition 8: *If f is integrable over $[a, M]$ for any $M > a$, then we define $\int_a^\infty f(x)\,dx$ to be $\lim_{M \to \infty} \int_a^M f(x)\,dx$, provided this limit exists. (For the definition of the limit of $f(x)$ as $x \to \infty$, the reader should see Exercise 9 of Section 5.1.)*

We leave it to the reader to define $\int_{-\infty}^a f(x)\,dx$.

Example 6: Consider

$$\int_1^\infty (1/x^2)\,dx. \tag{55}$$

Since $\int_1^M (1/x^2)\,dx = 1 - (1/M)$ and $\lim_{M\to\infty} (1 - (1/M)) = 1$, the value of (55) is 1.

Observe that in both Examples 5 and 6, the functions being integrated are bounded on the set over which the improper integrals are being taken ($(0, 1)$ in the case of $f(x) = x^2$ of Example 5, and $[1, \infty)$ in the case of $f(x) = 1/x^2$ of Example 6). We now give an example of an improper integral which converges even though the function is not bounded on the set over which it is being integrated.

Example 7:   Consider

$$\int_0^1 x^{-1/2}\,dx. \tag{56}$$

Then for any $p > 0$, we have $\int_p^1 x^{-1/2}\,dx = 2 - 2p^{1/2}$. The limit of $2 - 2p^{1/2}$ as $p \to 0$ is 2; therefore the value of (56) is 2. Observe, however, that the function $f(x) = x^{-1/2}$ is not bounded on $(0, 1]$.

Improper integrals have many of the same properties that ordinary Riemann integrals have. To prove propositions about improper integrals, one generally has to use Proposition 2 of Chapter 5. The following proposition gives some of the basic properties of improper integrals.

Proposition 26:   Suppose both of the improper integrals $\int_a^b f(x)\,dx$ and $\int_a^b g(x)\,dx$ (where a might be $-\infty$ and b might be ∞) converge. Then

a) $\int_a^b (f + g)(x)\,dx$ converges and is equal to $\int_a^b f(x)\,dx + \int_a^b g(x)\,dx$.
b) For any constant k, $\int_a^b (kf)(x)\,dx$ converges and is equal to $k \int_a^b f(x)\,dx$.
c) If $a < c < b$, then $\int_a^c f(x)\,dx$ and $\int_c^b f(x)\,dx$ both converge and

$$\int_a^b f(x)\,dx + \int_a^c f(x)\,dx = \int_c^b f(x)\,dx.$$

PROOF: We prove (a) and leave the proofs of (b) and (c) to the reader. There are two cases: (1) where one or both of the limits of integration are infinite and (2) where both limits of integration are finite. If both a and b are finite, then for any $p > 0$,

$$\int_{a+p}^{b-p} f(x)\,dx + \int_{a+p}^{b-p} g(x)\,dx = \int_{a+p}^{b-p} (f+g)(x)\,dx \tag{57}$$

(Proposition 12). Since the limit as $p \to 0$ of the left side of (57) exists and is equal to $\int_a^b f(x)\,dx + \int_a^b g(x)\,dx$ (Proposition 2 of Chapter 5), the limit of the right side of (57) exists and is equal to the same limit. Since the limit as $p \to 0$ of the right side of (57) is $\int_a^b (f+g)(x)\,dx$, we have proved (a) for the case that both a and b are finite. The case when one or both of the limits of integration are infinite is left as an exercise.

RIEMANN-STIELJES INTEGRAL

Let f be a function defined and bounded on $[a, b]$ and g be a function which is monotone increasing on $[a, b]$. Let P be any partition of $[a, b]$, and let M_i and m_i have the meaning assigned in Section 6.1. Then we can form the sums

$$U(P, f, g) = \sum_P M_i(g(x_i) - g(x_{i-1})) \tag{58}$$

and

$$L(P, f, g) = \sum_P m_i(g(x_i) - g(x_{i-1})). \tag{59}$$

Note that if $g(x) = x$, then $U(P, f, g)$ and $L(P, f, g)$ become the usual upper and lower Riemann sums of f over $[a, b]$ (see (2) and (3)). Because g is monotone increasing, we can prove the following proposition.

Proposition 27: If P' is a refinement of P, then

$$m(g(b) - g(a)) \leqslant L(P, f, g) \leqslant L(P', f, g) \leqslant U(P', f, g)$$
$$\leqslant U(P, f, g) \leqslant M(g(b) - g(a)), \tag{60}$$

where $m = \mathrm{glb}\{f(x) \mid x \in [a, b]\}$ and $M = \mathrm{lub}\{f(x) \mid x \in [a, b]\}$.

The proof of Proposition 27 is essentially a combination of the proofs of Propositions 1 and 2 with $g(x_{i+1}) - g(x_i)$ replacing $x_{i+1} - x_i$.

Proposition 27 in turn enables us to conclude that

$$\mathrm{glb}\{U(P, f, g) \mid P \text{ a partition of } [a, b]\} \tag{61}$$

and

$$\mathrm{lub}\{L(P, f, g) \mid P \text{ a partition of } [a, b]\} \tag{62}$$

both exist.

Definition 9: *We denote* (61) *by* $\overline{\int}_a^b f(x) \, dg(x)$ *and* (62) *by* $\underline{\int}_a^b f(x) \, dg(x)$. *If* $\overline{\int}_a^b f(x) \, dg(x) = \underline{\int}_a^b f(x) \, dg(x)$, *we denote the common value by* $\int_a^b f(x) \, dg(x)$. *We call* $\int_a^b f(x) \, dg(x)$, *if it exists, the* **Riemann-Stieljes integral of f with respect to g over $[a, b]$.** *If* $\int_a^b f(x) \, dg(x)$ *exists, we say that f is* **integrable with respect to g over $[a, b]$.**

Riemann-Stieljes integrals share almost all the properties of Riemann integrals; indeed, the proofs of most propositions concerning the Riemann-Stieljes integral are virtually identical to the proofs of the corresponding propositions about Riemann integrals, but with $g(x_{i+1}) - g(x_i)$ replacing $x_{i+1} - x_i$. For example, for Riemann-Stieljes integration, Proposition 3 becomes the following proposition.

Proposition 28:   The function f is integrable with respect to g on $[a, b]$ if and only if given any $p > 0$, there is a partition P such that

$$U(P, f, g) - L(P, f, g) < p.$$

Using Proposition 28 one can then prove the following proposition.

Proposition 29:   If f is continuous on $[a, b]$ and g is any monotone increasing function on $[a, b]$, then f is integrable with respect to g.

PROOF: Let $p > 0$ and p' be such that $p'(g(b) - g(a)) < p$. Since f is continuous on $[a, b]$, a compact set, f is uniformly continuous on $[a, b]$ (Proposition 31 of Chapter 3). Therefore there is $q > 0$ such that if t and t' are points of $[a, b]$ such that $|t - t'| < q$, then $|f(t) - f(t')| < p'$. Let P be any partition of $[a, b]$ with $m(P) < q$. Then for any subinterval $[x_i, x_{i+1}]$ determined by P, $M_i - m_i < p'$. Therefore

$$U(P, f, g) - L(P, f, g) = \sum_{i=0}^{n-1} (M_i - m_i)(g(x_{i+1}) - g(x_i))$$

$$< p'\left(\sum_{i=0}^{n-1} g(x_{i+1}) - g(x_i) \right) = p'(g(b) - g(a)) < p.$$

Therefore by Proposition 28, f is integrable with respect to g over $[a, b]$. (The reader should compare this proof with the proof of Proposition 5.)

In certain instances, the Riemann-Stieljes integral reduces to a Riemann integral.

Proposition 30:   If f is continuous on $[a, b]$ and g has a continuous derivative on $[a, b]$, then

$$\int_a^b f(x)\, dg(x) = \int_a^b f(x)g'(x)\, dx. \tag{63}$$

PROOF: Let $P = \{x_0, \ldots, x_n\}$ be any partition of $[a, b]$. Then for $i = 0, \ldots, n - 1$, it is possible to find w_i such that

$$g(x_{i+1}) - g(x_i) = g'(w_i)(x_{i+1} - x_i),$$

with $w_i \in (x_i, x_{i+1})$ (Proposition 17 of Chapter 5). Therefore

$$\sum_{i=0}^{n-1} f(w_i)(g(x_{i+1}) - g(x_i)) = \sum_{i=0}^{n-1} f(w_i)g'(w_i)(x_{i+1} - x_i). \tag{64}$$

As $m(P) \to 0$, the sides of (64) tend to the corresponding sides of (63).*

---

* This assumes that the sum on the right side of (64) converges to $\int_a^b f(x)\, dg(x)$ as $m(P) \to 0$. This is in fact the case; a proof can be obtained by simple adjustments of the techniques of Section 6.2.

The Riemann-Stieljes integral in certain instances reduces to a series (see Exercise 2). The various forms that the Riemann-Stieljes integral can assume (Riemann integral, series, and ordinary Riemann-Stieljes integral) sometimes allow certain ideas to be presented under the one heading of Riemann-Stieljes integral, whereas they would otherwise have to be considered separately.

EXERCISES

1. Prove that if f is integrable with respect to both g and h on $[a, b]$, then f is also integrable with respect to $g + h$ and $\int_a^b f(x) \, d(g + h)(x) = \int_a^b f(x) \, dg(x) + \int_a^b f(x) \, dh(x)$.

2. Let $I: \mathbb{R} \to \mathbb{R}$ be defined by $I(x) = 0$ if $x \leqslant 0$ and $I(x) = 1$ if $x > 0$.

a) Set $g(x) = I(x - c)$, where $a \leqslant c < b$. Prove that if f is continuous on $[a, b]$, then $\int_b^a f(x) \, dg(x) = f(c)$.

b) Suppose that $\sum a_n$ is a convergent series all of whose terms are positive and that $\{x_n\}$, $n \in \mathbb{N}$, is a sequence of distinct points of $[a, b)$. Set $g(x) = \sum_{n=1}^{\infty} a_n I(x - x_n)$, and suppose f is continuous on $[a, b]$. Prove that $\int_a^b f(x) \, dg(x) = \sum_{n=1}^{\infty} a_n f(x_n)$. Hint: Let $p > 0$. There is an integer M such that $\sum_{n=M}^{\infty} a_n < p$. Set

$$g_1(x) = \sum_{n=1}^{M-1} a_n I(x - x_n)$$

and

$$g_2(x) = g(x) - g_1(x).$$

By Proposition 1 and (a) (with finite induction), it follows that

$$\int_a^b f(x) \, dg(x) = \int_a^b f(x) \, dg_1(x) + \int_a^b f(x) \, dg_2(x)$$

$$= \sum_{n=1}^{M-1} a_n f(x_n) + \int_a^b f(x) \, dg_2(x).$$

Now prove that $\left| \int_a^b f(x) \, dg(x) - \sum_{n=1}^{M-1} c_n f(x_n) \right| < Kp$, where $K = \text{lub}\{|f(x)| \mid x \in [a, b]\}$.

3. Prove that f is integrable with respect to every monotone increasing function g on $[a, b]$ if and only if f is continuous on $[a, b]$.

4. Prove (b) and (c) of Proposition 26.

5. Suppose that $\sum_{i=1}^{\infty} a_i$, each $a_i \geqslant 0$, is a series, and f is a function such that $f(i) \geqslant a_i$ for all integers i greater than some positive integer M. Suppose too that f is integrable, non-negative, and monotone decreasing over $[M, b]$ for any $b > M$. Prove that $\sum_{i=1}^{\infty} a_i$ converges if $\int_M^{\infty} f(x)\, dx$ converges.

6. Prove that if f is integrable on every closed subinterval of (a, b) (a and b both finite) and if f is bounded on (a, b), then $\int_a^b f(x)\, dx$ converges.

7. Determine for what values of t $\int_0^{\infty} x^t\, dx$ converges.

7

SEQUENCES AND SERIES OF FUNCTIONS

7.1 TAYLOR'S THEOREM. POWER SERIES

Definition 1: *Let* $f^{(0)} = f$. *Then* $f^{(n)}$, *called the* **nth derivative** *of* f, *is defined to be the first derivative of* $f^{(n-1)}$, $n \geqslant 1$, *provided that* $f^{(n-1)}$ *is differentiable. We also call n the* **order** *of the derivative* $f^{(n)}$.

If $f^{(n)}$ *exists and is continuous on an interval* I, *then* f *is said to be of* **class** C^n *on* I. *If* f *has derivatives of all orders on* I, *that is, if* $f^{(n)}$ *exists on* I *for each positive integer n, then* f *is said to be of* **class** C^∞ *on* I.

Clearly, if f is of class C^n, $n \geqslant 1$, on some interval I, then f is also of class C^k, $0 \leqslant k \leqslant n$, on I.

Example 1: The function $f: \mathbb{R} \to \mathbb{R}$ defined by $f(x) = x^3$ is C^∞. The function $g: \mathbb{R} \to \mathbb{R}$ defined by

$$g(x) = \begin{cases} x^{n+1} & \text{if } 0 \leqslant x, \, n \text{ a positive integer,} \\ 0 & \text{if } x < 0 \end{cases}$$

is of class C^n, but not of class C^{n+1}.

Proposition 1 (*Taylor's Theorem*): Assume that f has derivatives of all orders up to $n+1$ on some open interval (a, b) and that $f^{(n)}$ is continuous on $[a, b]$. Then

$$f(b) = f(a) + f'(a)(b - a) + \cdots$$

$$+ \frac{f^{(n)}(a)}{n!} (b - a)^n + \frac{f^{(n+1)}(t')}{(n+1)!} (b - a)^{n+1}, \tag{1}$$

where t' is such that $a < t' < b$. (Recall that $0!$ is defined to be 1, and $n!$ is defined to be $n(n-1)(n-2) \ldots 2 \cdot 1$ for any positive integer n.)

PROOF: Set

$$F(x) = f(b) - \left(f(x) + f'(x)(b-x) + \ldots + \frac{f^{(n)}(x)}{n!}(b-x)^n \right)$$
$$+ Q(b-x)^{n+1}, \quad (2)$$

for all x in $[a, b]$, where

$$Q = \left[-f(b) + f(a) + f'(a)(b-a) + \ldots \right.$$
$$\left. + \frac{f^{(n)}(a)}{n!}(b-a)^n \right] \Big/ (b-a)^{n+1}. \quad (3)$$

It is easily verified that $F(a) = F(b) = 0$ and that F is differentiable on $[a, b]$. Consequently, by Proposition 17 of Chapter 5, there is t' in (a, b) such that $F'(t') = 0$. Therefore

$$F'(t') = (-f'(t') + f'(t')) + (f''(t')(b-t') - f''(t')(b-t')) + \ldots$$
$$+ f^{(n+1)}(t') /n!(b-t')^n - Q(n+1)(b-t')^n.$$

It follows that

$$Q = \frac{f^{(n+1)}(t')}{(n+1)!}. \quad (4)$$

Substituting the expression for Q in (4) in (3) and carrying out the proper algebraic steps, we find (1), which is what we wanted to prove.

Corollary 1: If f has derivatives of all orders up to $n+1$ on some open interval U which contains a, then given any $x \in U$

$$f(x) = \sum_{i=0}^{n} \frac{f^{(i)}(a)}{i!}(x-a)^i + \frac{f^{(n+1)}(t')}{(n+1)!}(a-x)^{n+1}, \quad (5)$$

where t' lies between a and x.

Corollary 2: If f is of class C^∞ on some open interval U which contains a, then for each $x \in U$ and each positive integer n, there is a number t'_n between x and a such that

$$f(x) = \sum_{i=0}^{n} \frac{f^{(i)}(a)}{i!}(x-a)^i + \frac{f^{(n+1)}(t'_n)}{(n+1)!}(a-x)^{n+1}. \quad (6)$$

The proofs of these corollaries are left as exercises.

Since $\sum_{i=0}^{n} (f^{(i)}(a)/i!)(x - a)^i$ is a polynomial function, Corollaries 1 and 2 tell us that certain functions can be "approximated" by polynomial functions at least over a certain interval. It would also appear that the more derivatives a function has, the better it can be approximated by polynomial functions.

Now let f be of class C^∞ on some open interval U which contains a and set

$$f_n(x) = \sum_{i=0}^{n} \frac{f^{(i)}(a)}{i!} (x - a)^i, \tag{7}$$

for each $x \in U$.

Then $\{f_n\}$, $n \in \mathbb{N}$, is a sequence of functions (in fact polynomial functions) defined and of class C^∞ on U. Moreover, if we compare (7) with the right side of (6), we would expect this sequence of functions to "converge" to f on U provided that

$$\frac{f^{(n+1)}(t'_n)}{(n + 1)!} (a - x)^{n+1} \to 0 \tag{8}$$

for each x in U. What we are really saying in such a case, however, is that the sequence of f_n will converge to f if given any x in U, the sequence $\{f_n(x)\}$, $n \in \mathbb{N}$, of real numbers converges to $f(x)$ in the usual sense of convergence of a sequence. Indeed, since we have no notion yet of what it means for a sequence of functions to converge, we will make this our definition.

SEQUENCES AND SERIES OF FUNCTIONS

Definition 2: *If $\{f_n\}$, $n \in \mathbb{N}$, is a sequence of functions defined on S, S any subset of $\mathbb{R}$, then we say that this sequence **converges** to a function f defined on S if for each $x \in S$, $f_n(x) \to f(x)$.* *If $\{f_n\}$, $n \in \mathbb{N}$, converges to f, we write $f_n \to f$ and say that f is the **limit** of $\{f_n\}$, $n \in \mathbb{N}$.*

Example 2: Let $f_n(x) = x^n$ for $x \in [0, 1]$. Then for $0 \leqslant x < 1$, $f_n(x) \to 0$, but for $x = 1$, $f_n(x) \to 1$. Therefore $\{f_n\}$, $n \subset \mathbb{N}$, converges to the function f defined by $f(x) = 0$, $x \neq 1$, and $f(1) = 1$ (defined on $[0, 1]$). Note that although each f_n is continuous (and even of class C^∞), f itself is discontinuous at 1. We might therefore ask: Under what conditions does a sequence of continuous functions converge to a continuous function?

---

* We have not formally defined what is meant by a sequence of functions on S. By analogy with the definition of a sequence of real numbers, however, it is simply a function from $\mathbb{N}$ into the set of functions from S into $\mathbb{R}$, that is, a correspondence with each positive integer n of a function $f_n : S \to \mathbb{R}$.

Example 3:   According to the way we have defined convergence for a sequence of functions, the functions f_n defined in (7) converge to f if and only if condition (8) applies. Therefore it is reasonable to ask: When can we be sure that condition (8) applies? We also note that (7) is the partial sum of a series; consequently, (1) the techniques developed for dealing with series may be of some help in our investigations in this chapter and (2) if $f_n \to f$, where f_n is as defined in (7), then it is justifiable to write

$$f(x) = \sum_{i=0}^{\infty} \frac{f^{(i)}(a)}{i!} (x - a)^i. \tag{9}$$

Definition 3:   *The right side of (9) is called **Taylor series expansion** of f about a. In the special case that $a = 0$, we sometimes call the Taylor series expansion of f about 0 the **Maclaurin series** for f.*
    *Given a sequence $\{f_n\}$, $n \in \mathbb{N}$, each $f_n : S \to \mathbb{R}$, we define the **series based on** $\{f_n\}$, $n \in \mathbb{N}$, to be the sequence*

$$f_1, \; f_1 + f_2, \; f_1 + f_2 + f_3, \ldots,$$

*and call $\sum_{i=1}^{n} f_i$ the **nth partial sum** of the series; f_i is the **ith term** of the series.*
*Since a series of functions is but a special type of sequence of functions, the notion of convergence as defined in Definition 2 can be applied to series of functions. If a series of functions converges to a function f, we call f the **sum** of the series and the series is said to **converge**. If the series does not converge, it is said to **diverge**.*

We may denote the series based on $\{f_n\}$, $n \in \mathbb{N}$, by $\sum_{i=1}^{\infty} f_i$.
    *If $f_n(x) = a_n x^n$, or $f_n(x) = a_n(x - a)^n$ for each $x \in S$, then $\sum_{i=0}^{\infty} f_i$ is said to be a **power series**.*

A Taylor series or a Maclaurin series is a power series. Although there are important series other than power series, most of our discussion will be about power series. We begin by looking at the power series

$$\sum_{i=0}^{\infty} a_i x^i = a_0 + a_1 x + a_2 x^2 + \ldots \tag{10}$$

and finding for what values of x the power series in (10) converges.

Proposition 2:   The series in (10) either converges for all values of x, only for $x = 0$, or there is a positive number K such that if $|x| < K$, then (10) converges absolutely, but if $|x| > K$, then (10) diverges.

PROOF: Assume that (10) converges for $x = b$. Suppose $|x| < |b|$. Then it follows that

$$\sum_{i=0}^{\infty} |a_i x^i| = \sum_{i=0}^{\infty} |a_i b^i| \, |x/b|^i. \tag{11}$$

Since $\sum a_i b^i$ converges, $a_n b^n \to 0$; consequently there is $M > 0$ such that $|a_i b^i| < M$ for $i = 0, 1, 2, \ldots$. It follows from (11) then that

$$\sum_{i=0}^{\infty} |a_i x^i| \leqslant M \sum_{i=0}^{\infty} |x/b|^i. \tag{12}$$

Since $\sum_{i=0}^{\infty} |x/b|^i$ is a geometric series with ratio less than 1, it converges. Therefore $\sum_{i=0}^{\infty} |a_i x^i|$ converges by the Comparison Test; hence (10) converges absolutely. Letting $K = \text{lub}\left(b \, \bigg| \, \sum_{i=0}^{\infty} a_i b^i \text{ converges}\right)$, we obtain the K called for in Proposition 2. If the set of b for which (10) converges has no upper bound, then (10) converges for all values of x.

INTERVALS OF CONVERGENCE

Definition 4: *If $K > 0$ is such that (10) converges (absolutely) for $|x| < K$ (that is, for $x \in (-K, K)$), but diverges for $|x| > K$, then we call K the **radius of convergence** and $(-K, K)$ the **interval of convergence** of (10).*

(As we shall see, a power series may or may not converge at the end points of its interval of convergence.)

We now turn our attention to finding the radius of convergence of $\sum a_i x^i$. One way in which K, the radius of convergence for $\sum a_i x^i$, can be found is by using the Ratio Test (Proposition 13 of Chapter 4). Consider

$$\left| \frac{a_{n+1} x^{n+1}}{a_n x^n} \right| = |a_{n+1}/a_n| \, |x|. \tag{13}$$

If $\sum a_i x^i$ is to converge absolutely, it suffices that there be $r < 1$ and an integer M such that $n > M$ implies (13) is less than r. Let $K' = \text{lub}\{x \mid \text{there is } r < 1 \text{ for which there is an integer } M \text{ such that } n > M \text{ implies (13) is less than } r\}$. Then we have the following proposition.

Proposition 3: The number K' (as defined in the preceding paragraph) is less than or equal to the radius of convergence of $\sum a_i x^i$. That is, $\sum a_i x^i$ will converge absolutely if $|x| < K'$.

Usually, K' will also be the radius of convergence. In particular, we have the following proposition.

Proposition 4:   If K' is as defined and if $|x| > K'$ implies that there is $r > 1$ such that there is an integer M for which $n > M$ implies (13) is greater than r; then K' is the radius of convergence.

PROOF: In this instance, $\sum |a_i x^i|$ will converge for $|x| < K'$, but will diverge if $|x| > K'$ (Proposition 13 of Chapter 4). Since $\sum a_i x^i$ converges absolutely only for x inside its interval of convergence, the interval of convergence must be $(-K', K')$; hence K' is the radius of convergence.

We now conclude this section with several examples.

Example 4:   Consider $\sum_{i=1}^{\infty} x^i$. Then $|x^{i+1}/x^i| = |x|$. If $|x| < 1$, then $\sum_{i=1}^{\infty} x^i$ will converge absolutely.   But if $|x| > 1$, then the series $\sum |x^i|$ will diverge. The interval of convergence for $\sum x^i$ is therefore $(-1, 1)$. We note that the series diverges at both end points of its interval of convergence. Such is not necessarily the case as we will see from the next example.   We also note that $\sum x^i$ is a geometric series with first term x and ratio x (hence we have another way of telling that its interval of convergence is $(-1, 1)$ since such a series will converge if and only if the absolute value of its ratio, here $|x|$, is less than 1). Using the formula for the sum of a geometric series, we find $\sum_{i=1}^{\infty} x^i$ converges to $x/(1 - x)$ if $x \in (-1, 1)$. Thus we see that $\sum_{i=1}^{\infty} x^i$ converges to the function $f: (-1, 1) \to \mathbb{R}$, where $f(x) = x/(1 - x)$.

Example 5:   Consider

$$\sum_{i=1}^{\infty} (x - 1)^i/i. \tag{14}$$

Although (14) is not a power series of the form (10), it can be handled using the same techniques which apply to (10).   For if we set $y = x - 1$, then (14) takes the form

$$\sum_{i=1}^{\infty} (1/i) y^i. \tag{15}$$

If we find the interval of convergence for (15), say $(-K, K)$, then (14) will converge absolutely if and only if $|x - 1| < K$. The radius of convergence for (14) will still be K, but the *interval of convergence* for (14), that is, the interval on which (14) converges absolutely, will be centered about 1 instead of 0.   Now

$$|(y^{i+1}/(i + 1))/(y^i/i)| = \left(\frac{i}{i + 1}\right)|y|. \tag{16}$$

Since $n/(n + 1) \to 1$, the interval of convergence for (15) will be $(-1, 1)$. Therefore the interval of convergence for (14) will be $(-1 + 1, 1 + 1) = (0, 2)$. We now consider if (14) converges at the end points of $(0, 2)$, that is, for $x = 0$ and $x = 2$. For $x = 0$, (14) becomes

$$\sum_{i=1}^{\infty} (-1)^i/i, \tag{17}$$

which converges; while if $x = 2$, (14) becomes $\sum_{i=1}^{\infty} (1/i)$, which does not converge.

Example 6: Consider

$$\sum_{i=0}^{\infty} x^i/i!. \tag{18}$$

Now $|(x^{i+1}/(i+1)!)/(x^i/i!)| = (1/(i+1)) |x|$. Since $(1/(i+1)) |x| \to 0 < 1$ for any value of x, we see that (18) converges absolutely for every real number x.

EXERCISES

1. Prove that the Maclaurin series expansion for any polynomial function f is f itself.

2. Prove that if f is a function which is of class C^{∞} over the entire set of real numbers and if $u \leqslant f^{(n)} \leqslant U$, u, U real numbers, for all $n > M$, M some positive integer, then given any real number a, the Taylor series expansion for f about a converges to f for any real number x.

3. Prove Corollaries 1 and 2 of Proposition 1.

4. Find the interval of convergence for each of the following series. Test the series to see if they converge at the end points of the interval of convergence.

a) $\sum_{i=1}^{\infty} (-1)^i x^i/(i+1)$

b) $\sum_{i=1}^{\infty} x^{i-1}/i^3$

c) $\sum_{i=0}^{\infty} (-1)^i (x-4)^{i+1}/(2i+1)!$

d) $\sum_{i=0}^{\infty} (-1)^i (x-10)^i/9^i$

5. Consider the power series $\sum_{i=1}^{\infty} a_i x^i$ and $\sum_{i=1}^{\infty} b_i x^i$.

a) Prove that $\sum a_i x^i \sum b_i x^i$ converges absolutely for any x in the intervals of convergence of both series. (For the definition of the product of two series, see Section 4.3.)

b) Prove that if $\sum a_i x^i$ and $\sum b_i x^i$ converge to the functions f and g on some open interval U, then $\sum a_i x^i \sum b_i x^i$ converges to fg on U.

6. Determine the limits (if they exist) of the sequences of functions defined in each of the following. Each function will be defined on $[0, 1]$.

a) $f_n(x) = x/n$
b) $f_n(x) = (x + 1)^{-n}$

c) $f(x) = \begin{cases} 0 & \text{if } x = 0 \\ x^{-n} & \text{if } x \neq 0 \end{cases}$

7.2 UNIFORM CONVERGENCE

We have seen that a sequence of continuous functions need not converge to a continuous function (Example 2). We now introduce a concept which enables us to give a partial answer to the question: When does a sequence of continuous functions converge to a continuous function?

Definition 5: *Suppose $\{f_n\}$, $n \in \mathbb{N}$, is a sequence of functions, all defined on W. Then we say that $\{f_n\}$, $n \in \mathbb{N}$, **converges uniformly** to the function g on W if given any $p > 0$, there is an integer M such for every $x \in W$ and $n > M$, $|g(x) - f_n(x)| < p$.*

Definition 5 tells us that $\{f_n\}$, $n \in \mathbb{N}$, converges uniformly to g on W if we not only have $f_n(x) \to g(x)$ for each $x \in W$—that is, for any $p > 0$ and any $x \in W$, there is M (which depends on x and p) such that $n > M$ implies $|f_n(x) - g(x)| < p$—but given any $p > 0$, there is an integer M such that for any $x \in W$, $|f_n(x) - g(x)| < p$ (that is, there is an M which depends only on p and not on both p and x).

Proposition 5: If $\{f_n\}$, $n \in \mathbb{N}$, converges uniformly to the function g on the set W, then if each f_n is continuous at $a \in W$ (at least for n sufficiently large), then g is continuous at a.

PROOF: Let $p > 0$. Then there is an integer M such that $n > M$ implies $|f_n(x) - g(x)| < p/3$ for all $x \in W$. Let $n > M$. Since f_n is continuous at a there is $q > 0$ such that $|x - a| < q$ implies $|f_n(x) - f_n(a)| < p/3$. Suppose then we have $x \in W$ with $|x - a| < q$. Then $|g(x) - g(a)| \leqslant |g(x) - f_n(x)| + |f_n(x) - f_n(a)| + |f_n(a) - g(a)| < p/3 + p/3 + p/3 = p$. Therefore g is continuous at a.

Example 7: The sequence of functions in Example 2 converges to a discontinuous function, thus the convergence could not be uniform. We may, however, have a sequence of continuous functions which converges to a continuous function even though the convergence is not uniform. For example, let $g_n: \mathbb{R} \to \mathbb{R}$ be defined by $g_n(x) = x/n$ (the graph of g_n is simply the line with slope $1/n$). Then $g_n \to h$, where $h(x) = 0$ for all $x \in \mathbb{R}$. This convergence follows from the fact that $g_n(x) = x/n \to 0$ for all $x \in \mathbb{R}$. Nevertheless, this convergence is not uniform on $\mathbb{R}$; for given any $p > 0$ and any positive integer n, one can find x sufficiently large so that $|g_n(x) - h(x)| = |x/n - 0| = x/n > p$. We have then that $g_n \to h$, h a continuous function, even though the convergence is not uniform.

Suppose K is any compact subset of $\mathbb{R}$ (for example, K might be a closed interval). Then $g_n \to h$ uniformly on K; we show this as follows. Let k be the largest absolute value assumed by any $x \in K$ (how do we know there is such a largest absolute value if K is non-empty?). Choose any $p > 0$ and let M be a positive integer large enough so that $k/M < p$. Then for $n > M$ we have $|g_n(x) - h(x)| = |x/n| < k/M < p$ for $x \in K$. Therefore the convergence is uniform on K.

We see from the next two propositions that limits of uniformly convergent sequences of functions may share other important properties with the functions of the sequence besides continuity.

Proposition 6: If $\{f_n\}$, $n \in \mathbb{N}$, converges uniformly to g on $[a, b]$ and each f_n is integrable over $[a, b]$, then g is also integrable over $[a, b]$. Moreover, $\int_a^b f_n(x)\,dx \to \int_a^b g(x)\,dx$.

PROOF: Let $p > 0$. Choose $q > 0$ such that $q(b - a) < p/3$. Since $f^n \to g$ uniformly on $[a, b]$, there is some integer n for which

$$|f_n(x) - g(x)| < q, \quad \text{for all } x \in [a, b]. \tag{19}$$

Since each f_n is integrable over $[a, b]$, there is a partition P of $[a, b]$ such that

$$U(P, f_n) - L(P, f_n) < p/3. \tag{20}$$

Now (19) implies that $g(x) < f_n(x) + q$ on $[a, b]$. It follows then (the details are left as an exercise) that

$$U(P, g) \leqslant U(P, f_n) + p/3. \tag{21}$$

Similarly, since $g(x) \geqslant f_n(x) - q$ (from (19)), we also have

$$L(P, f) \geqslant L(P, f_n) - p/3. \tag{22}$$

We therefore have

$$U(P, g) - L(P, g)$$
$$= (U(P, g) - U(P, f_n)) + (U(P, f_n) - L(P, f_n) + (L(P, f_n) - L(P, g))$$
$$< p/3 + p/3 + p/3 = p.$$

Therefore g is integrable over $[a, b]$ by Proposition 3 of Chapter 6.

Now choose an integer M such that $n > M$ implies

$$|f_n(x) - g(x)| < p/(b - a) \quad \text{for all } x \in [a, b]. \tag{23}$$

Then if $n > M$, we have $|\int_a^b g(x) \, dx - \int_a^b f_n(x) \, dx| = |\int_a^b (g - f_n)(x) \, dx| \leq \int_a^b |g - f_n| (x) \, dx$ (Proposition 14 of Chapter 6) $< (b - a)(p/(b - a)) = p$. Therefore $\int_a^b f_n(x) \, dx \to \int_a^b g(x) \, dx$.

Corollary: If $f_n \to g$ uniformly on $[a, b]$, if each f_n is of class C^1 (that is, each f_n has a continuous derivative on $[a, b]$), and if $f'_n \to h$ uniformly on $[a, b]$, then g is of class C^1 on $[a, b]$ and $g' = h$.

PROOF: Since $f'_n \to h$ uniformly and each f'_n is continuous on $[a, b]$, h is also continuous on $[a, b]$. If x is any point of $[a, b]$, then $f'_n \to h$ uniformly on $[a, x]$. Therefore by Proposition 6 we have

$$\int_a^x f'_n(t) \, dt \to \int_a^x h(t) \, dt. \tag{24}$$

By the Fundamental Theorem of the Calculus and the fact that f'_n is the first derivative of f_n, we have $\int_a^x f'_n(t) \, dt = f_n(x) - f_n(a)$. Consequently, from (19) we obtain

$$(f_n(x) - f_n(a)) \to \int_a^x h(t) \, dt. \tag{25}$$

But $f_n(x) \to g(x)$ and $f_n(a) \to g(a)$; hence (25) implies

$$\int_a^x h(t) \, dt = g(x) - g(a) \quad \text{for any } x \in [a, b]. \tag{26}$$

It follows from Proposition 21 of Chapter 6 that $h = g'$ on $[a, b]$. Since h is continuous, g is of class C^1 on $[a, b]$.

Example 8: Consider once again the sequence of functions in Example 2. Since $\int_0^1 f_n(x) \, dx = \int_0^1 x^n \, dx = 1/(n + 1) \to 0 = \int_0^1 f(x) \, dx$, we have the conclusion of Proposition 6, even though the f_n do not converge to f uniformly.

Example 9: Let $f_n : \mathbb{R} \to \mathbb{R}$ be defined by $f_n(x) = (\sin(nx))/\sqrt{n}$. Then $f_n \to f$, where $f(x) = 0$ for all $x \in \mathbb{R}$. However, $f'_n(x) = \sqrt{n}(\cos nx)$. Since $f'(x) = 0$ for all $x \in \mathbb{R}$, $f'_n \nrightarrow f'$. For example, $\{f'_n(0) = \sqrt{n}\}, n \in \mathbb{N}$, does not converge at all. It follows then that the f_n do not converge uniformly to f even on the compact subsets of $\mathbb{R}$.

We prove now another criterion for uniform convergence.

Proposition 7: The sequence of functions $\{f_n\}$, $n \in \mathbb{N}$, defined on W, converges uniformly to some limit g on W if and only if given any $p > 0$ there is an integer M such that whenever n and m are integers greater than M, then

$$|f_n(x) - f_m(x)| < p \quad \text{for all } x \in W. \tag{27}$$

PROOF: Suppose first that $f_n \to g$ uniformly on W. Then for some integer M, $n > M$ implies $|f_n(x) - g(x)| < p/2$ for all $x \in W$. Therefore if m and n are both larger than M, we have

$$|f_n(x) - f_m(x)| \leqslant |f_n(x) - g(x)| + |g(x) - f_m(x)| < p/2 + p/2 = p$$

for all $x \in W$.

Suppose now that there is an integer M such that $n, m > M$ implies (27). Then for each $x \in W$, $\{f_n(x)\}$, $n \in \mathbb{N}$, is a Cauchy sequence and, hence, converges to some limit which we will call $g(x)$. Thus $f_n \to g$ on W. It remains to be shown that this convergence is uniform.

Suppose $p > 0$ and M is such that $n, m > M$ implies (27). Now

$$|f_n(x) - g(x)| \leqslant |f_n(x) - f_m(x)| + |f_m(x) - g(x)|. \tag{28}$$

Since $f_m(x) \to g(x)$, the right side of (28) will ultimately be less than p. This, in turn, implies that for $n > M$, $|f_n(x) - g(x)| < p$ for all $x \in W$. Consequently, $f_n \to g$ uniformly on W.

Proposition 8: Suppose the power series $\sum a_i x^i$ has $(-K, K)$ as its interval of convergence and $[a, b] \subseteq (-K, K)$. Then $\sum a_i x^i$ converges uniformly on $[a, b]$.

PROOF: If we let F_n denote the nth partial sum of $\sum a_i x^i$, then by Proposition 7 we will show that $\sum a_i x^i$ converges uniformly on $[a, b]$ if we show that given any $p > 0$ there is an integer M such that for $M < n < m$, $|F_n(x) - F_m(x)| < p$ for all $x \in [a, b]$. Now

$$|F_n(x) - F_m(x)| = \left| \sum_{i=n}^{m} a_i x^i \right| \leqslant \sum_{i=n}^{m} |a_i x^i| \leqslant \sum_{i=n}^{m} |a_i z^i|, \tag{29}$$

where z is that end point of $[a, b]$ at which each $|a_i| |z|^i$ has its maximum value. (It is left to the reader to prove (1) the maximum of each $|a_i| |x|^i$ on $[a, b]$ occurs at an end point of $[a, b]$ and (2) it will be the same end point of $[a, b]$ for each $|a_i| |x|^i$.) Since $\sum |a_i z^i|$ converges, given any $p > 0$, there is an integer M such that $M < n < m$ implies $\sum_{i=n}^{m} |a_i z^i| < p$. This, in turn, implies $|F_n(x) - F_m(x)| < p$; hence $\sum a_i x^i$ converges uniformly on $[a, b]$.

Corollary: A power series converges to a continuous function on its interval of convergence.

PROOF: Suppose $x \in (-K, K)$, where K is the radius of convergence of the power series $\sum a_i x^i$. Then there is a positive number p such that $x \in (-K + p, K - p)$ and $[-K + p, K - p] \subseteq (-K, K)$. Since $[-K + p, K - p]$ is a closed subinterval of $(-K, K)$, $\sum a_i x^i$ converges to a function which is continuous at x by Proposition 5. Since x was an arbitrary point of $(-K, K)$, $\sum a_i x^i$ converges to a continuous function on all of $(-K, K)$.

Proposition 8 can be generalized to the following: A power series converges uniformly on any compact subset of its interval of convergence. (Note that a closed interval is but one particular type of compact set.) We leave the proof of this generalization to the reader.

Not only is each partial sum of a power series a continuous function, it is a class C^∞ function. Given the power series $\sum a_i x^i$, the "derivative series" of $\sum a_i x^i$ (that is, the sequence of the derivatives of the partial sums of $\sum a_i x^i$) is

$$\sum i a_i x^{i-1}, \tag{30}$$

which is itself a power series. The corollary to Proposition 6 then gives us the following proposition.

Proposition 9: If $\sum a_i x^i$ converges uniformly to g on $[a, b]$ and if $\sum i a_i x^{i-1}$ converges uniformly to h on $[a, b]$, then g is of class C^1 on $[a, b]$ and $g' = h$. (That is, we can find the power series for the derivative of g on $[a, b]$ by taking the "derivative" of $\sum a_i x^i$ provided that $\sum i a_i x^{i-1}$ converges uniformly on $[a, b]$.)

We now answer the question: Where does (30) converge uniformly?

Proposition 10: Suppose the interval of convergence for $\sum a_i x^i$ is $(-K, K)$. Then $(-K, K)$ is also the interval of convergence for (30).

PROOF: Let K' be the radius of convergence for (30); we must show that $K = K'$. Suppose that (30) converges absolutely for some x. Then $(1/|x|)(\sum |i a_i x^i| = \sum |i a_i x^{i-1}|$; hence $\sum |i a_i x^i|$ also converges. But $|a_i x^i| \leq |i a_i x^i|$; hence $\sum a_i x^i$ converges absolutely by the Comparison Test. Consequently, $K' \leq K$.

Suppose now that $\sum a_i x^i$ converges absolutely for x and $0 < |y| < |x|$. Then

$$\sum i |a_i y^{i-1}| = (1/|y|) \sum i |a_i x^i| \, |y/x|^i. \tag{31}$$

Since $\sum |a_i x^i|$ converges, there is a number M such that $|a_i x^i| < M$ for each i.

Therefore the right side of (31) is less than or equal to

$$M/(|y| \sum i \, |y/x|^i). \tag{32}$$

Since $y/x < 1$, the Ratio Test proves that $\sum i |y/x|^i$ converges. Therefore $\sum i |a_i y^{i-1}|$ is a bounded monotone increasing sequence and, hence, converges. We therefore conclude that $K \leqslant K'$. Hence $K = K'$.

Since (30) is itself a power series, it converges uniformly on $[-K + p,$ $K + p]$ for any sufficiently small $p > 0$, where $(-K, K)$ is its interval of convergence which is the same as the interval of convergence for $\sum a_i x^i$. It follows that one can take the derivative of (30) at any point of $(-K, K)$ to obtain the second derivative of $\sum a_i x^i$; this second derivative is $\sum i(i-1)a_i x^{i-2}$ (with $i - 2 \geqslant 0$), and it will have the same interval of convergence as $\sum a_i x^i$. By finite induction we can prove the following proposition.

Proposition 11: The power series $\sum a_i x^i$ converges to a function g of class C^∞ on its interval of convergence. Moreover, the nth derivative series of $\sum a_i x^i$, that is, the series of the nth derivatives of the partial sums of $\sum a_i x^i$, converges to $g^{(n)}$ with the same interval of convergence as $\sum a_i x^i$.

One can also prove—we leave the proof as an exercise for the reader—the following proposition.

Proposition 12: If $\sum a_i x^i$ converges to the function g on $(-K, K)$, then $a_i = g^{(i)}(0)/i!$ (that is, $\sum a_i x^i$ is the Taylor series expansion of g about 0).

Note Proposition 12 implies the uniqueness of a power series expansion; that is, if $\sum a_i x^i$ and $\sum b_i x^i$ are both power series expansions of f, then $a_i = b_i$ for all i.

Since a power series converges uniformly on any closed subinterval of its interval of convergence, Proposition 6 gives us the following result.

Proposition 13: If $\sum a_i x^i$ converges to g on $(-K, K)$ and $[a, b] \subseteq (-K, K)$, then $\int_a^b g(x) \, dx = \int_a^b (\sum a_i x^i) \, dx = \sum \int_a^b a_i x^i \, dx$.

EXERCISES

1. Prove (21) in the proof of Proposition 6.

2. Prove that part of the proof of Proposition 8 which was left as an exercise for the reader.

3. Prove Proposition 12.

4. Prove directly that the sequence of functions in Example 2 does not converge uniformly on $[0, 1]$.

5. Prove that if $\{f_n\}$, $n \in \mathbb{N}$, is a sequence of functions defined on W such that $|f_n(x)| \leqslant K_n$ for each $n \in \mathbb{N}$ and $x \in W$, then $\sum\limits_{i=1} f_i$ converges uniformly on W if $\sum K_i$ converges.

6. Consider the sequence of functions defined by

$$f_n(x) = x/(1 + nx^2), \quad x \in (-1, 1).$$

Prove that $f_n \to g$ uniformly on $(-1, 1)$ (what is g?) and $f'_n(x) \to g'(x)$ for $x \neq 0$, but $f'_n(0) \nrightarrow g'(0)$. What hypotheses of the corollary to Proposition 6 are not satisfied here?

7. Prove that $\sum a_i x^i$ and $\sum (a_i/(i + 1))x^{i+1}$ have the same intervals of convergence.

8. Restate several of the propositions we have proved about a power series of the form $\sum a_i x^i$ to apply to power series of the form $\sum a_i(x - a)^i$. We have really lost no generality in our results by restricting our attention to series of the form $\sum a_i x^i$.

7.3 SOME SPECIAL FUNCTIONS

LOGARITHMIC AND EXPONENTIAL FUNCTIONS

Definition 6: *Let $x > 0$. We define the **natural logarithm** of x by*

$$\int_1^x (1/t) \, dt. \tag{33}$$

Since $f(t) = 1/t$ is continuous on any closed interval of positive numbers, the natural logarithm of any positive number exists. We will denote the natural logarithm of $x > 0$ by $L(x)$.

The following proposition gives some immediate consequences of the definition of $L(x)$.

Proposition 14

a) $L(1) = 0$.
b) $L(x) > 0$ if $x > 1$.
c) $L(x) < 0$ if $0 < x < 1$.
d) $L'(x)$ exists for any $x > 0$ and is equal to $1/x$.

PROOF: (a) and (b) are clear. (c) follows from the fact that $0 < x < 1$ implies $L(x) = \int_1^x (1/t)\, dt = -\int_x^1 (1/t)\, dt$, and $\int_x^1 (1/t)\, dt$ is positive. (d) follows from Proposition 21 of Chapter 6.

The following proposition gives further properties we expect of logarithms.

Proposition 15: Let a and b be positive real numbers. Then

$$L(ab) = L(a) + L(b).$$

PROOF: Set $f(x) = L(ax)$. Then $f'(x) = L'(ax)(a)$ (by the Chain Rule) $= a(1/ax) = 1/x = L'(x)$. Since $f'(x) = L'(x)$, $f(x) = L(x) + k$ (by the corollary to Proposition 18 of Section 5.3), for some constant k. Letting $x = 1$, we find $f(1) = L(a) = L(1) + k = 0 + k = k$. Therefore

$$L(x) + L(a) = L(ax). \tag{34}$$

Letting $x = b$, (34) gives $L(a) + L(b) = L(ab)$.

Since the derivative of L is continuous and positive for all positive real numbers, we can prove the following proposition.

Proposition 16: L is monotone increasing and one-one on the set of positive real numbers. (Cf. Exercise 6, Section 5.3.)

Proposition 17: Given any real number a, there is a positive real number b such that $L(b) = a$; that is, L is onto $\mathbb{R}$.

PROOF: Suppose $0 < a$. We already know $L(1) = 0 < a$. We now prove there is a positive integer n such that $a < L(n)$. For any integer $n > 1$, consider the partition P of $[1, n]$ defined by $\{1, 2, \ldots, n - 1, n\}$. Since $1/t$ is monotone decreasing for positive t, we have

$$L(n) = \int_1^n (1/t)\, dt > \sum_{i=2}^n (1/i)(1) = \sum_{i=2}^n (1/i).$$

Since $\sum_{i=2}^n (1/i)$ fails to converge as n goes to infinity, there must be n such that $L(n) > a$. Since $0 < a < L(n)$ and L is continuous, by the Intermediate Value Theorem there exists a b (with $1 < b < n$) such that $L(b) = a$.

We leave the case $a < 0$ as an exercise for the reader.

Corollary: We have then that L is one-one, onto $\mathbb{R}$, differentiable at each point of its domain, and monotone increasing. We therefore conclude that $L^{-1} : \mathbb{R} \to \mathbb{R}$ is a function which is defined, continuous, and differentiable at each real number x, and the range of L^{-1} is the set of positive real numbers.

PROOF: The conclusions follow at once from Proposition 37 of Chapter 3 and Proposition 13 of Chapter 5.

Definition 7:   *We define the inverse of L to be the **exponential function**. We denote the exponential function by E. We denote $E(1)$ by e and define e^x to be $E(x)$.*

The next proposition gives some of the basic properties of the exponential function.

Proposition 18

 a) $E(x)E(y) = E(x + y)$ for any real numbers x and y.
 b) $E(x)/E(y) = E(x - y)$.

(Note that according to the notation introduced in Definition 7, Proposition 18 states the familiar laws of exponents $e^x e^y = e^{x+y}$ and $e^x/e^y = e^{x-y}$.)

PROOF: We prove (a) and leave the proof of (b) to the reader. We have $L(E(x)E(y)) = L(E(x)) + L(E(y)) = x + y = L(E(x + y))$. Since L is one-one, it follows that $E(x)E(y) = E(x + y)$.

Proposition 19

 a) For any $a \in \mathbb{R}$, $E'(a) = E(a)$.
 b) $E(0) = 1$.
 c) The Maclaurin series expansion for E is $\sum_{i=0}^{\infty} x^i/i!$. This series converges absolutely to $E(x)$ for any $x \in \mathbb{R}$.

PROOF: (a) According to Proposition 13 of Chapter 5, if $L(b) = a$, then $E'(a) = 1/L'(b) = 1/(1/b) = b$. But if $L(b) = a$, then $b = E(a)$ (since $E = L^{-1}$). Therefore $E'(a) = E(a)$.
Statement (b) follows at once from the fact that $L(1) = 0$. We leave (c) to the reader.

Using Definition 7 and Proposition 18 one can prove that if q is a rational number (that is, if $q = m/n$, m and n integers, $n \neq 0$), then $(E(x))^q = (e^x)^q = e^{qx}$ for any $x \in \mathbb{R}$. We extend this result by defining $(e^x)^y = e^{xy}$ for any x, $y \in \mathbb{R}$.

We can now use the logarithmic and exponential functions to define powers and logarithms relative to numbers other than e.

Definition 8:   *Suppose $a > 0$. Then $L(a)$ exists and $a = E(L(a)) = e^{L(a)}$. For any $x \in \mathbb{R}$, set $a^x = (e^{L(a)})^x = e^{xL(a)} = E(xL(a))$.*

If $L(a) \neq 0$, that is, if $a > 0$, $a \neq 1$, then $f: \mathbb{R} \to \mathbb{R}$ defined by $f(x) = a^x$ is onto the set of positive real numbers: moreover, f is one-one, differentiable, and monotone (increasing if $a > 1$, and decreasing if $0 < a < 1$). Therefore, f^{-1} is a function defined, continuous, and differentiable on the set of positive real numbers. If $b > 0$, we define $f^{-1}(b)$ to be the *logarithm to the base a of b*, denoted by $\log_a b$. By definition then $a^{\log_a b} = b$.

Proposition 20

(a) If a and b are positive numbers, $a \neq 1$, then $\log_a b = L(b)/L(a)$. From which it follows that if f is a function from the positive real numbers into $\mathbb{R}$ defined by $f(x) = \log_a x$, then $f'(x) = 1/(xL(a))$.

(b) For $a > 0$ and $x \in \mathbb{R}$, $L(a^x) = xL(a)$.

(c) If $f: \mathbb{R} \to \mathbb{R}$ is defined by $f(x) = a^x$, $a > 0$, then $f'(x) = a^x L(a)$.

(d) If $a > 0$, $a \neq 1$, $b > 0$, and $x, y \in \mathbb{R}$, we have $(a^x)^y = a^{xy}$ and $\log_a(b^x) = x \log_a b$.

PROOF: (a) We have $b = e^{L(b)} = a^{\log_a b} = e^{L(a)\log_a b}$. Since E is one-one, we also have $L(b) = L(a)\log_a b$, from which we obtain $\log_a b = L(b)/L(a)$. If $f(x) = \log_a x = L(x)/L(a)$, then $f'(x) = L'(x)/L(a) = 1/xL(a)$.

(b) $L(a^x) = L(e^{xL(a)}) = xL(a)$ (since $L = E^{-1}$).

Part (c) follows at once from the Chain Rule and the fact that $a^x = e^{xL(a)}$. We leave the proof of (d) as an exercise.

Considerably more might be said about the logarithmic and exponential functions as well as the number e. Since we have confirmed many of the most important properties, techniques for handling other problems are implied in what we have done. We illustrate this point in the next example.

Example 10: Set $f(x) = x^x$. Since $x = e^{L(x)}$, $f(x) = e^{xL(x)}$. Because $L(x)$ is only defined for $x > 0$, $f(x)$ is defined only for $x > 0$. We may therefore consider f thus defined to be a function from the set of positive real numbers into $\mathbb{R}$. For each x for which f is defined, f is also differentiable. Using the Chain Rule we find $f'(x) = e^{xL(x)}(L(x) + x/x) = x^x(L(x) + 1)$ for $x > 0$.

THE SINE AND COSINE FUNCTIONS

It is assumed that the reader has already seen a definition of the sine and cosine using the unit circle and radian angle measurement. As an alternate approach we will define the sine and cosine as functions whose derivatives have certain properties.

Definition 9: *We define the **sine function,** denoted by S, to be that function (if one exists and is unique) which has the properties $S''(x) = -S(x)$ for each real number x and $S(0) = 0$ and $S'(0) = 1$.*

*We define the **cosine function,** denoted by C, to be that function (if one exists and is unique) for which* $C''(x) = -C(x)$ *for all* $x \in \mathbb{R}$ *and with* $C(0) = 1$ *and* $C'(0) = 0$.

Since $S''(x) = -S(x)$, we have $S^{(3)}(x) = -S'(x)$ and $S^{(4)}(x) = -S''(x) = -(-S(x)) = S(x)$. From this we have the following proposition.

Proposition 21:   If the sine function exists, it will be of class C^{∞}. Moreover, we have $S^{(n)}(x) = S^{(n+4)}(x)$, $n = 0, 1, 2, 3 \ldots$

Since $S(0) = 0$, $S'(0) = 1$, $S''(0) = 0$, and $S^{(3)}(0) = -1$, we have the following corollary.

Corollary: If the function S exists, its Maclaurin series expansion must be

$$\sum_{i=0}^{\infty} (-1)^i (x^{2i+1}/(2i+1)!) = x - x^3/3! + x^5/5! - x^7/7! + \ldots \quad (35)$$

Since (35) is found to converge absolutely for all real numbers, $S(x)$ is defined for all $x \in \mathbb{R}$.

(We could in fact have used (35) as the definition of $S(x)$.)

Applying similar reasoning to the cosine function, we have this proposition.

Proposition 22:   The cosine function is defined and of class C^{∞} on all of $\mathbb{R}$; its Maclaurin series expansion is

$$\sum_{i=0}^{\infty} (-1)^i (x^{2i}/(2i)!) = 1 - x^2/2! + x^4/4! - \ldots . \quad (36)$$

Using the series representations of the sine and cosine, we can prove the following proposition.

Proposition 23

a) $S(-x) = -S(x)$ and $C(-x) = C(x)$ for all $x \in \mathbb{R}$.
b) $S'(x) = C(x)$ and $C'(x) = -S(x)$ for all $x \in \mathbb{R}$.

It is also quite possible to prove basic trigonometric identities using our approach to the sine and cosine. For example, we have the following proposition.

Proposition 24

$S^2(x) + C^2(x) = 1$ for all $x \in \mathbb{R}$. Hence $|S(x)|$ and $|C(x)|$ are both at most 1 for all $x \in \mathbb{R}$.

PROOF: Set $f(x) = S^2(x) + C^2(x)$. Then $f'(x) = 2S(x)S'(x) + 2C(x)C'(x) = 2S(x)C(x) - 2S(x)C(x) = 0$. Therefore f is a constant function. For $x = 0$, $f(0) = S^2(0) + C^2(0) = 1$. Therefore $f(x) = 1 = S^2(x) + C^2(x)$ for all $x \in \mathbb{R}$.

Once the basic properties of the sine and cosine are found, the rest of the trigonometric functions can be defined and their properties developed.

EXAMPLES OF FUNCTIONS WITH SPECIAL PROPERTIES

Example 11: (A differentiable function with a discontinuous first derivative): We revert to the more usual notation of $\sin x$ and $\cos x$ to denote the sine and cosine of x, respectively. Define $f: \mathbb{R} \to \mathbb{R}$ by

$$f(x) = \begin{cases} x^2 \sin(1/x) & \text{if } x \neq 0, \\ 0 & \text{if } x = 0. \end{cases}$$

If $x \neq 0$, then $f'(x) = 2x \sin(1/x) - \cos(1/x)$. We now compute $f'(0)$.

$$f'(0) = \lim_{x \to 0} \frac{f(x) - f(0)}{x - 0} = \lim_{x \to 0} \frac{x^2 \sin(1/x)}{x} = \lim_{x \to 0} x \sin(1/x). \qquad (37)$$

Since $|\sin(1/x)| \leqslant 1$ and $x \to 0$, the last limit in (37) is 0. Therefore $f'(0) = 0$. However, $\lim_{x \to 0} f'(x)$ does not exist; therefore f' is not continuous at 0.

Example 12: (A function which is integrable on $[0, 1]$, but which is discontinuous at every rational number in $[0, 1]$): The series $\sum_{i=1}^{\infty} (1/2)^i$ is absolutely convergent with sum 1; hence any rearrangement of this series also sums to 1. Since the rational numbers in $[0, 1]$ are countable, there is a one-one correspondence of these numbers with the full set of positive integers. Let t be such a correspondence and let t_m denote that rational number of $[0, 1]$ which corresponds to the positive integer m. Define the function f as follows: For each $x \in [0, 1]$, set

$$f(x) = \sum_{t_j \leqslant x} (1/2)^j. \qquad (38)$$

At each rational number t_m, f will have a "jump" of $(1/2)^m$; hence f will be discontinuous at t_m. Moreover, f is bounded below by 0 and above by 1 (we will have $f(1) = 1$). The function f, however, is also monotone increasing on $[0, 1]$; hence f is integrable on $[0, 1]$ (Proposition 6 of Chapter 6).

Example 13: (A function f which is C^∞ on all of $\mathbb{R}$, but whose Maclaurin series expansion converges to $f(x)$ only when $x = 0$): Let $f(x) = e^{-1/x^2}$ if $x \neq 0$ and set $f(0) = 0$. For any $x \neq 0$ f is the composition of E, the exponential function, with h, where $h(x) = -1/x^2$. Since E and h are C^∞ when $x \neq 0$, f is C^∞ when $x \neq 0$. We also claim that f has derivatives of all orders at 0 and that the value of $f^{(n)}(0)$ is 0 for any positive integer n. This later fact follows (with some work) from the fact that

$$\lim_{x \to 0} e^{-1/x^2}/x^n = 0 \quad \text{for any } n \in \mathbb{N}. \tag{39}$$

We prove (39) as follows. Set $y = 1/x^2$. Then $\lim_{x \to 0} e^{-1/x^2}/x^n = \lim_{y \to \infty} y^{n/2}/e^y = 0$ (by L'Hôpital's Rule). The Maclaurin series expansion of f then is identically 0, while $f(x)$ is 0 only when $x = 0$. Therefore the Maclaurin series expansion of f represents f only at 0 even though f is C^∞ on all of $\mathbb{R}$.

We have already seen an example of a function which is continuous on all of $\mathbb{R}$ but which is not differentiable on all of $\mathbb{R}$, namely, $f(x) = |x|$. It is possible to find a function which is continuous on all of $\mathbb{R}$ but which is not differentiable anywhere. We will not construct such a function in this text. The interested reader can find an example in Gelbaum and Olmsted, *Counterexamples in Analysis*, Holden-Day Publishing Co., or in most of the standard analysis texts.

EXERCISES

1. Prove Proposition 16.

2. Prove (c) of Proposition 19.

3. Prove (d) of Proposition 20.

4. Use the function in Example 12 to prove that given any positive integer n, there is a function f defined on $(-1, 1)$ such that f has derivatives of all orders up to and including n on $(-1, 1)$, but $f^{(n)}$ is discontinuous at 0.

5. Suppose $a = 0$, $a \neq 1$, and $f: \mathbb{R} \to \mathbb{R}$ is defined by $f(x) = a^x$.

a) Show that f is one-one and onto the set of positive real numbers.

b) Show that f is monotone decreasing if $a < 1$ and monotone increasing if $1 < a$.

6. Find the set of real numbers x for which the following expressions make sense. Indicate where the functions defined by these expressions are differentiable and find the first derivative in each case.

a) $f(x) = 3^x$

b) $f(x) = \log_5 x$

c) $f(x) = \log_3 \sin x$

d) $f(x) = \cos(\ln x^2)$ *

e) $f(x) = \log_4(5^x)$

f) $f(x) = e^{\log_5(x^3-1)}$

7. Prove that there cannot be a function $f : \mathbb{R} \to \mathbb{R}$ which is continuous only at each rational number.

8. Prove that the series in (35) and (36) converge for each $x \in \mathbb{R}$.

---

* $\ln x^2 = \log_e x^2 = L(x^2)$.

INDEX OF SYMBOLS

INDEX